The Land Is Full

Nelson Byrd Woltz
Landscape Architects

The Land Is Full

Nelson Byrd Woltz Landscape Architects

ESSAYS BY
ROBERT POGUE HARRISON
BRENT LEGGS
NINA-MARIE LISTER
THOMAS L. WOLTZ

EDITED BY
BRADFORD MCKEE

FOREWORD BY
ANDREA WULF

FOREWORD
Andrea Wulf
06

I

THE LAND IS FULL
Thomas L. Woltz
12

22 Centennial Park
Nashville, Tennessee

36 Bok Tower Gardens
Lake Wales, Florida

56 Hudson Yards Public Square and Garden
New York, New York

II

LANDSCAPES OF CARE: DESIGNING FOR AN ECOLOGY OF EMPATHY
Nina-Marie Lister
76

84 Brooklyn Naval Cemetery Landscape
Brooklyn, New York

96 Duke University Reclamation Pond
Durham, North Carolina

108 The Ramble
Fayetteville, Arkansas

III

A PURPOSEFUL PURSUIT OF EQUITABLE INTERPRETATION
Brent Leggs
128

IV

THE SAVING POWER
Robert Pogue Harrison
182

CREDITS
246

134 Alethia Tanner Park
Washington, D.C.

146 Machicomoco State Park
Gloucester County, Virginia

164 Georgia Institute of Technology EcoCommons
Atlanta, Georgia

188 Flight 93 National Memorial
Shanksville, Pennsylvania

202 Aga Khan Garden
Edmonton, Alberta

218 Memorial Park
Houston, Texas

FOREWORD

ANDREA WULF

We are on a mission; we are called to educate the world.
—Novalis, 1797

The ancient Greek word *poietikos* means creative or productive—and of course it's the root of our term poetry. "To poeticize is to create," explained the young German poet Novalis in the last decade of the eighteenth century, and he demanded that the "sciences must all be poeticized." For him and the Young Romantics, poetry could be anything from a poem and a novel to a painting, a building, a piece of music, or even a scientific experiment. For me, an NBW-designed landscape belongs to this list too. I always have to think of Novalis's ambition to unite the arts and the sciences, as well as humankind and nature, when I see an NBW project. It's all there: rigorous scientific research, an investigation of cultural and historical meaning, a philosophy of beauty—all built onto a deep-rooted belief that we're part of nature. This is what NBW calls landscape architecture, but I believe it's so much more.

For many years now, I have been in a prolonged conversation with Thomas Woltz, brainstorming a better term for this profession. To a layperson, a landscape architect is, all too often, simply a maker of pretty gardens. But is there a better term? During those years, we've failed to find one. And maybe we have to accept that landscape architecture at its best can't be defined in one or two words.

Instead, I will use as many words as it takes to describe what NBW does so magnificently.

They are

finders of stories, shapers of land;

listeners of silent voices;
restorers of ecologies,
makers of beauty,

helpers of animals,
translators of memories;

tellers of truths;

capturers of wonders,
facilitators of contemplation,
creators of joy,
revealers of layers;

dancers of lines,
surveyors of places,
carers of plants;

conductors of movements;
keepers of dreams;

healers of ecosystems,
explorers of archives,
poets of grasslands;

painters of imagination;

rescuers of tortoises,
preservers of authenticity,
connectors of people,
protectors of wetlands;

musicians of shapes;

interrogators of sites,
stewards of the land;

. . . to me, they are the most
glorious triangulators of
nature, history, and beauty.

I

ESSAY 12 THE LAND IS FULL
Thomas L. Woltz

PROJECTS 22 CENTENNIAL PARK
Nashville, Tennessee

36 BOK TOWER GARDENS
Lake Wales, Florida

56 HUDSON YARDS PUBLIC SQUARE AND GARDEN
New York, New York

ESSAY

THE LAND IS FULL

Thomas L. Woltz

All land, urban or rural, wild or toxic, comes with a deep history. As landscape architects, when we first approach any site, it is with a simple act of grace: we ask its story and then begin to find the imprints left by the forces that have shaped it over time. We call this *slow listening*. It is the way we gain the knowledge that inspires and shapes the vision for that land. It allows us to open our hearts and minds and develop a sense of kinship with the place. This approach has kept us from regarding any land as empty or vacant or lacking in value or history. The land precedes us and will survive us. It holds eons in layered and complex mysteries. The land is full. It has no end of stories to tell us if we are patient enough to listen.

Approaching land with this sense of its fullness shifts the ways we engage with it in our daily lives as landscape architects and as human beings. It is the counter position to phrases referring to land as "tabula rasa," "blank slate," or even the seemingly benign "green space." As designers and researchers, we draw inspiration from the complex of geological forces, ecology, and animal and human agency that has shaped the land. It presents tantalizing threads that lead us to design inspiration that arises from the site. We have established two essential categories of interrogation: the physical ecology of the land and the layers of culture that evolve with that ecology over time. Building an understanding of the intimate connections among humans and the ecologies they inhabit allows us to compose a portrait of a place. Knowing the place is the first act of design.

The biography of a site originates in that land's formation. It includes volcanic activity, glaciation, plate tectonics and seismic activity, erosion, and sedimentation. The mineral content and dynamics of what earth scientists call the "parent" geology establish the base of the chemical and physical qualities of the soil, which in turn hosts specific plant and animal communities adapted to the site's pH, microbiome, tilth, and soil fertility—what is generally called ecology. The dynamics of water's movement across land, the forestation, climate patterns, and fire further form the unique ecological nuance of a site. In a historic continuum, these conditions have provided specific factors that influence patterns of animal and human migration. Over time these movements determine settlement patterns around corridors favorable to the broad exploitation and reallocation of natural resources. It is remarkable how often the prehistoric geology of a place, the richness of its soil, a specific forest type, the power of falling water, or the rhythms of flooding are at the crux of the "how" and "why" of populations, settlements, and modes of land use. These revelations of the ecology-to-culture continuum are what we seek through a profound commitment to research.

A Research-Driven Design Process

At Nelson Byrd Woltz Landscape Architects we have developed a research-based design process that invests substantial resources in learning the deep history of a site. The firm's staffing structure mirrors our research-based design process: imagine a celestial model of a sun with three revolving planets. The Design Group, our sun, is at the core of what we do. Orbiting this team are the Conservation, Culture, and Communication Groups, our planets, like tides and gravity, an intimate relationship between entities. These groups push and shape the design approach that occupies their common center by revealing and interpreting the unique aspects of a site early in every design process. They carry within them the specialized perspectives and methods of their fields.

The tools of the Conservation Group are broad and include recorded data from national and regional resources such as the U.S. Geological Survey's geologic models and soil maps. Often we engage geotechnical engineers to augment this information for greater specificity about a site's soils and structural integrity through borings that reveal the underlying geology. Biological assessments performed on a given site over days, weeks, and even years provide the crucial next layers of information. The assessment teams are often led by the NBW Conservation Group's specialists in restoration ecology and conservation biology and will include, at times, research scientists who specialize in such areas as amphibians and reptiles, birds, bats, insects, native and nonnative plants, terrestrial mammals, and even lichens. These ecological assessments give us glimpses into the biological balance and health of a site and often inspire the direction of the design team in restoring what might have thrived on a given site and might thrive once again. They also tell us of any specific micro-ecologies to be protected or conserved on a site. Not least, they inform the long-term management and maintenance that follows the implementation of a design. Newly restored landscapes require intensive management to become established as healthy ecosystems, resilient against shifts in climate and the introduction of multitudes of aggressive nonnative vegetation and animals.

Parallel to the ecological survey, the Culture Group explores archives, databases, insurance maps, deeds, and other document trails for clues to the site's recorded past. The richest sources are primary and personal, such as oral histories, wills, diaries, and letters that reveal unique knowledge and memories of persons with intimate attachments to a place. The stories of many sites come alive through legacies told in families and communities over generations that find their repository among individuals alive today. There are always multiple families and multiple histories to consider between a site's earliest human presence and its current occupation. Historians, anthropologists, and other scholars buttress our knowledge of the site through publications and direct contributions. Archaeologists employ powerful tools available for their surveying practices, such as Light Detection and Ranging (LIDAR) technology, which enables high-resolution imaging of ground elevation with exquisite accuracy and without disturbance and can reveal features or phenomena in layers of land cover that can be isolated for visual analysis. The resulting images capture anomalies within the soil profile that may reveal previous excavations made for foundations, cultivation, earthworks, or graves. Together these tools provide the design team with a rich body of information, the fingerprints of the site, for an intimate understanding of the land-language over time—stories known and unknown, revealed, and hidden. With this documentation the Culture Group builds the unique timeline of a site that captures the specific dynamics of human occupation and engagement with the land over thousands of years. Patterns emerge from this information as we collate it with ecological findings. We develop the narrative into a formal report, which follows the rigorous standards of a Cultural Landscape Report (CLR), a format established by the National Park Service, which oversees official designations of historically significant lands and structures within the U.S. Department of the Interior. The Park Service's format is applicable in the U.S. and other countries for documenting site history, contextual analysis, and the disposition or alterations to elements of the landscape. These extensive reports identify what are known among historians as "periods of significance," which are eras or moments in times past that were most important in its human occupation—and perhaps the end of its occupation. These periods are unique to the site under study; they may be transient, sudden, or prolonged over decades or even centuries, and they relate to the evolution of the site in ways that are crucial in grounding appropriate design adaptations we make in the present and are projecting into the future.

We explore the land's past for the communities we serve as landscape architects—our clients, the project stakeholders, the knowledge keepers. This is a first step in a conversation with the community. We listen to their passions, memories, and hopes about the site, knowing that design can only move at the speed of trust with a community. We get to know the people whom a landscape serves, so that any remaking of the site is useful to them over a long span of time. This is how we build stewardship, by creating authentic places that honor the past and that belong to the people who live in that place.

As the design of a landscape evolves, the Communications Group, our third planet, engages the complexity of sites to help people understand the continuum of the land's history. The group's tools—words, films, graphics, and digital models—communicate and clarify the design concepts. This graphic collateral provides the images for signage and creative concepts of wayfinding and interpretation on the site, telling the story that grew from the story of the land.

Unexpected Truths

The land is a witness that allows us to remember. When land is lost, we lose that witness, and without a witness, people can start to question whether historic events happened or not. Our research process often reveals painful site histories that have been covered or erased. We work to tell the truth about what lies within the land, because sustaining the land is sustaining the truth. It's essential to reckoning with legacies of indigeneity, subjugation, hate, injustice, violence. The more precise

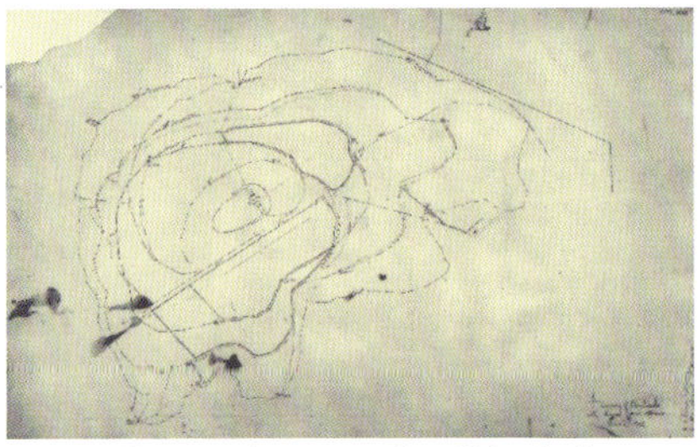

fig. 1 The 1809 insurance plat of Monticello.

fig. 2 George Willis Jr. (left) walking to his car followed by Maddox, pointing a short pistol, and his son holding an ax handle, July 4, 1964.

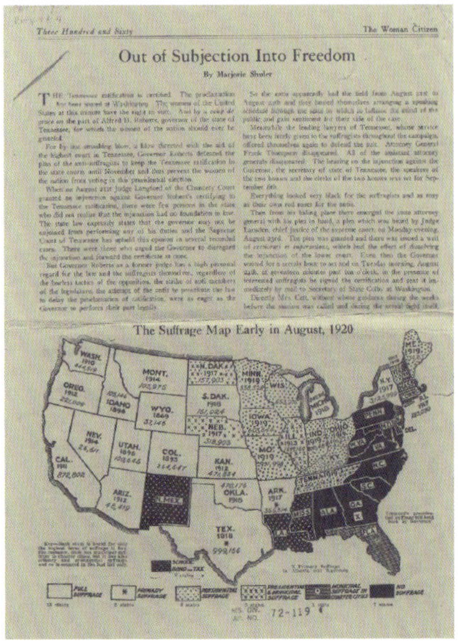

fig. 3 Map of the range of suffrage across states in early August, 1920.

the story and the location, the more power the designed landscape holds to reach our imaginations and our spirits and to inspire dialogue and eventual healing. The opportunity to protect these stories through the design of sites of memory and contemplation has offered potent ways to move a diverse public—people both injured by these realities or unaware of them—toward confrontation of the difficult aspects of our shared history.

Among our recent projects are the rejuvenation of the Burial Ground for Enslaved People and the creation of the Contemplative Site that places into the landscape the names of the 607 people Thomas Jefferson held in bondage during his lifetime at Monticello. This Burial Ground was brought to light by oral histories of descendants who knew this hallowed place in the forest, neither marked nor documented, as the burial place of their ancestors. More than twenty-five years of direct engagement of dozens of descendants of enslaved persons at Monticello, conducted by the Getting Word Project at the Thomas Jefferson Foundation, informed the design and material decisions during the restoration of the Burial Ground. The descendants were critical informants toward creating a place of dignity, solace, and reflection. (1)

Other truths have revealed themselves unexpectedly, as happened during research for the EcoCommons at Georgia Tech University in Atlanta. The project was steering toward returning the biodiversity of the Piedmont region to a large new campus open space. A review of past property documents showed that The Pickrick, a restaurant, once occupied a part of the site. Before its demolition, The Pickrick became a place of crucial importance to the Civil Rights story of the nation.

On July 3, 1964, one day after the Civil Rights Act became law to prohibit racial segregation, three students from the Interdenominational Theological Center, George Willis Jr., Woodrow Lewis, and Albert L. Dunn, placed their bodies in danger to test whether the new law would be upheld. After violently being forced out, the students were represented by the attorney Constance Baker Motley, who took their case to a three-judge circuit court in Atlanta. The federal court upheld the Civil Rights Act's prohibitions on racial exclusions in public accommodations. The restaurant's owner, Lester Garfield Maddox Sr., a future Georgia governor, closed the restaurant rather than abide by the law prohibiting him from continuing to racially discriminate against patrons. Upon learning this history, NBW developed a section of the site to articulate a narrative of the incident at The Pickrick, which was not originally a part of the project. The design finds its power in reviving the exact outline of The Pickrick in a space now called Unity Plaza, un-erasing it around a figurative retelling of the students' courage. (2)

In Centennial Park, in Nashville, the firm's research showed that the site's attractions to humans through time owes to a confluence of geological, hydrological, and ecological forces. Nearer to our own time, that confluence, through so many inevitabilities, made Centennial Park the site of the landmark rally that prompted ratification of the 19th Amendment to guarantee women the right to vote. (3) The design marks the route of the historic march in 1920 by suffragettes who

fig. 4 The Tennessee State Chamber at the moment the historic vote on national suffrage was counted on the floor.

fig. 5 An uncovered shell midden at Machicomoco State Park.

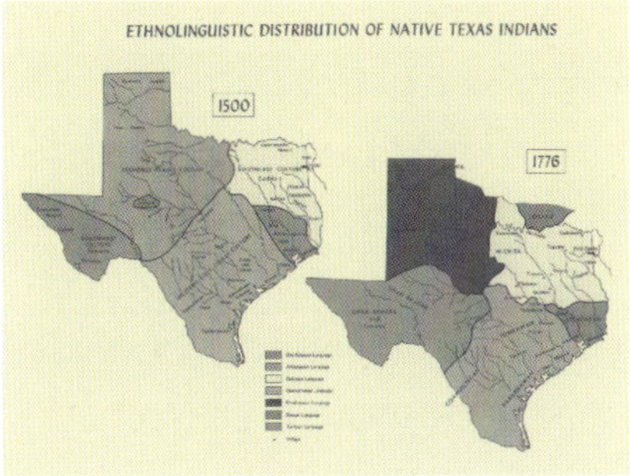

fig. 6 The ethnolinguistic distribution of native Texas Indians from 1500 and 1776.

demanded that Tennessee state legislators approve the amendment. Their protest, which took them to the steps of the Parthenon within the park, succeeded as the tipping point for Tennessee to become the 36th state to ratify women's voting rights. We worked closely with the Nashville sculptor Alan LeQuire and the Tennessee Woman Suffrage Monument organization to contextualize the bronze statues and commemorate this crucial moment for the state and the nation. The figures are sited at a level that allows visitors to conceptually march with the powerful women immortalized in the space. The five women honored in the new monument were present during the final ratification battle in 1920: Anne Dallas Dudley of Nashville; Abby Crawford Milton of Chattanooga; J. Frankie Pierce of Nashville; Sue Shelton White of Jackson; and Carrie Chapman Catt of New York. [4]

The existence of these landscapes and the design vision for each of them emerged from deep site interrogation to construct a spatial context to hold their stories, to inform visitors with their full narratives, while aspiring to bring comfort to those confronting regularly, or for the first time, difficult, complex, or painful histories.

Research to prepare us for design has also revealed unexpected ecological histories that have been lost over time owing to evolving land management practices. Our findings have come in the course of examining the seed bank (dormant seeds held in soil), investigating soil horizons (the visible layers of contrasting material layered deep in soil), or in reviewing written records of historic conditions. [5] In Memorial Park in Houston, six-foot-deep soil borings revealed parallel layers of ash, showing that, as long as 400 years ago, the Karankawa and Akokisa societies managed the landscape with fire, likely to attract grazing animals, including bison, to the banks of the Buffalo Bayou. In the contemporary design of Memorial Park, NBW designed both wet and dry prairie ecosystems to restore both hydric and fire-dependent and fire-tolerant grassland ecologies to the contemporary landscape. [6]

A seed bank in New Zealand, at a coastal site on the North Island, showed pre-settlement conditions of temperate rainforest on what later became severely overgrazed nonnative pasture in the twentieth century. The design team worked closely with elders and young leadership from the local Maori community (Ngā Tamānuhiri) to establish a nursery for germinating and cultivating the seeds of native trees that were later installed among the 600,000 trees planted to date, restoring seven miles of coastline to the rainforest conditions that would have existed before the first Maori landing, around 1300. At Bok Tower Gardens in Lake Wales, Florida, the presence of gopher tortoises was the clue that their habitat, sandhill ecosystem, severely compromised by development and agricultural practices in the region, could be restored to a pre-settlement, fire-managed ecology dominated by wiregrass and saw palmetto. This would better support the health of the gopher tortoise and the historic ecological community. Each of these examples of restoration ecology arose from biological assessments and curiosity about the complex, deep ecology of each site. Today, these flourishing ecosystems are managed to support their survival in a future of rapidly changing climate.

fig. 7 A comparison of the "Beautiful" and the "Picturesque" in landscape gardening from AJ Downing's *A Treatise on the Theory and Practice of Landscape Gardening*, 6th edition, 1860.

From Findings to Form

Form, in the design of landscapes, creates movement, memory, and experience. It also offers a scaffold for connecting people with the fullness of the land and its stories. The synthesis of the foundational research into ecology and culture inspires the early design forms, which often emerge as a gesture or composition against which the natural setting and ecology becomes more legible—a cultural datum for the unfolding of the dynamics of time and ecology. A purposeful tension between the context and the designed intervention sparks an awareness in people and engages them in nature. The forms may be as simple as a line, an angle, a circle, or an ellipse. These simple forms provide an intuitive orientation that frees the mind and the eye to absorb the setting and offers lines against which a person can read the subtlety and complexity of the land. There is a moral imperative to landscape architecture in deepening our connection to nature.

This imperative raises the question of what is natural. The past two centuries of landscape architecture have yielded a spectrum of answers. At the extremes there are highly constructed evocations of "nature untouched" and complete abstractions and subjugations of natural processes by the architecturalizing of plantings, landforms, and pavements. These extremes of expressing nature are equal in their artifice. The meandering paths, exuberant vegetation, and weeping conifers typical of Picturesque Period landscapes are as contrived as the abstract landscape patterning of postmodern designs, where concrete paths and mown lawns create a highly controlled public experience in nature. (7) Despite the heavily composed nature of both extremes, one is commonly construed as natural and the other as artificial. Even landscapes that seem natural—wild, romantic, sinuous—are ordered and maintained. Bringing aesthetics and ethics together, we believe form provides the structure that facilitates awareness of the designed landscape acting as the foil for measuring and monitoring the natural dynamics of a site. Awareness breeds stewardship, and it benefits the landscape and the communities vested in it to understand that all landscapes are constructs of some order, creations that require care. The image of nature untouched by humans has been shattered by numerous recent studies and publications, notably Charles C. Mann's book *1491*, which examines the North American landscape and its maintenance by Indigenous people over thousands of years. Nature is part of us, and we are a part of nature.

Each of the public landscapes shown in this book has relied upon the research-driven design process to discover a language of forms that elevates the experience of the site. The process of design renders that language as earthworks and paths entangled with introduced horticulture and restored ecosystems. These design compositions reside in dialogue with dynamic ecological processes that shape and steer the composition over time. Each unique composition is neither pure patternmaking nor pure restoration ecology but uses constructed form to reveal stories, shape the public experience, and heighten awareness of

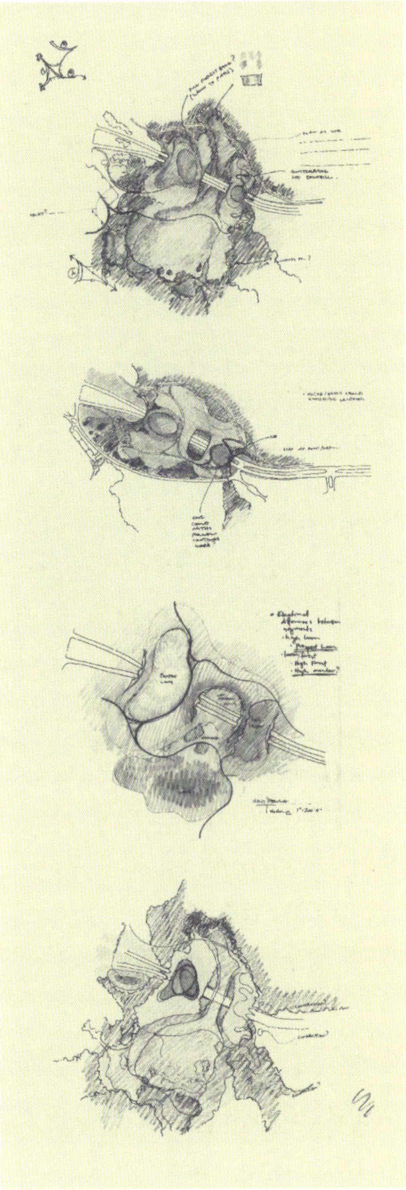

fig. 8 Hand-drawn sketches explore potential designs for the Memorial Park land bridge and prairie.

people to the ecological dynamics of the landscape. The design allows people to stand upon the very spot of a historic event, or in the shade of a tree that has been growing for four centuries, or to touch the unearthed spring water used by Indigenous people and settlers. These are the authentic moments when we can all hold hands with history and feel the connection to land's past in our bodies. The rectangle (slightly obscured by an exuberant meadow) of Unity Plaza on Georgia Tech University's campus recalls the exact location of The Pickrick, the civil-rights site that was made invisible by its demolition. A line through Machicomoco State Park creates a proportional time scale that holds thousands of years of Algonquian history and points to the resilience and future of Virginia Indians. The massive ellipse of the Eastern Glades of Memorial Park is an intuitive orientation device that leads people to a wide range of regional ecologies: forested wetland, native oak motts, prairie, and a lake, all held in balance by one geometry, which was inspired by an early design concept for the park from a century ago. For all else form does, it also quite simply helps us find our way.

Composition and Time in the Design Process

The extensive set of finely calibrated instruments we carry within our bodies are constantly, automatically gathering data, remembering details, and logging impressions while simply walking a site. When immersed in a landscape, the body unconsciously measures distances, slopes, textures, and space. Over days and seasons, observing light, temperature, humidity, vistas, sounds, and, perhaps most powerfully, the smells of a site, we can metabolize a site into our own bodies, and it becomes part of our personal memory bank. Extensive immersion uniquely positions the designer to have an empathic relationship to a given site and prepares the creative mind to hold and then to reveal what is true in the land. [8]

Once the fundamental ecological and cultural research is known by the design team, when the voices of the user and the stewards have been heard, when the body has spent time walking the site in various conditions, finally it is time to draw. This is the most critical moment in the birth of a creative design concept. And drawing is one of the most powerful tools designers have as another kind of physical engagement that triangulates among the eye, the hand, and the land. Design concepts improve dramatically through iteration, repetition, and testing. Whether we work in graphite, ink, watercolor, or clay, we are using our bodies to represent the existing conditions we discover and to propose new interventions in a site, converting these metabolic inputs into a creative output fueled by knowing the land. Creative inspiration in our design process is not, therefore, the result of a lightning strike, but rather the product of a mind and body well prepared over time in the making of a specific landscape. During this period of fecund creativity, the gestural forms, scales, and spaces of the design emerge simultaneously with the materials and realities of experience, the building materials, horticulture, water, and light.

Paramount to our work is imagining the visitor's experience both upon completion of construction as well as a century into the future, weighting the future experience over the immediate. It is a uniquely frustrating quality of landscape architecture that a design is least impactful the day it is completed; only with time will it evolve into a living system and achieve the complexity of the long-term vision. Infant landscapes are fragile, require dedicated care, and are vulnerable to the forces of climate extremes, vandalism, and neglect. There is a unique sense of time and patience given the decades that landscapes take to mature and achieve ecological stability. We work on a generational time scale. We work to inspire the stewardship of land and to ensure access to nature for people we will never know, to build places we are not likely to see mature in our lifetimes. This deep awareness is not a source of sadness or anxiety; landscape as a construct beyond ourselves and our human timeline is a source of great personal calm and perspective: landscape patience.

Preservation and the Middle of History

Land is in peril the world over. Once devastated, whether by war, neglect, or development, we lose the ecologies and stories of that land, and we become broken from our past; we lose the witness of our human history. Over the past two and a half decades, NBW has had the honor and opportunity to work with mindful and spiritual people who have challenged and expanded our sense of time, ability to see land, and the very definition of stewardship. These valuable lessons have become manifest in our process and worldview. We see ourselves and our work somewhere in the middle of a site's history—responding to its past and thinking of its far future. Time is not linear, and it moves in slippery ways, in loops and spirals; the past is not over, and our intervention is not necessarily a new chapter, nor is it an end of anything. This concept of time has shaped our sense of a covenant with the land and with those who tend it by seeing ourselves at a midpoint in the continuum “of history. An ever-shifting midpoint.

Covenant with Land

The land allows us to remember who we are and have been. The landscapes presented in this book are the manifestation of a moral bond with land held by a community of makers: a commitment to the preservation of its health, its history, and its future. Our mission is to conceive of, design, and build landscapes that forge a meaningful connection between people and nature. We do this creative work not by magic or stroke of genius, but through a steady process of deep listening to learn the land, the people who use it, and those who tend it. It is hard work, often invisible work, requiring patience and vision. However, nothing is more gratifying to me than seeing a severely damaged ecology rebounding, a suppressed history revealed, retold, and preserved, or a family or friends exploring a vast public landscape where restored wildlife and plant life thrive once again in their ancient environs.

There is much work to do because the land is full.
We thrive as humans only because the land is full.
We will persevere in this work because the land is full.

MANY CHAPTERS LATER

PROJECT
LOCATION

CENTENNIAL PARK
NASHVILLE, TENNESSEE

Centennial Park began as a fairground for the 1897 Tennessee Centennial and International Exposition. When NBW began work in 2013, the park had a faded Beaux-Arts core made for fairground buildings that no longer exist and one, the Parthenon, that does. Around the core are long informal greenswards of mature trees in Picturesque arrangements. The park needed a functional plan and a more engaging identity to serve three million people a year who arrive alone or by the thousands. NBW was asked to elevate the visitor experience, not least to favor public health and fitness, and modernize its dated infrastructure.

The project began with research into the history, human and not, of the site and its surroundings. In some ways, the park seemed stuck in 1903, when it became Nashville's first public park several years after the Exposition. The six-month fair celebrated the state's hundred years of perseverance and aimed to bring a "great revelation to the world of the industrial resources of Tennessee," as one opening-day speaker promised. The exhibits presented the state's minerals and timber. It rolled out the latest locomotive, the finest farm equipment, and "the flying machine of the future," among other assets. The exhibition halls were built as neoclassical palaces, following the model of the 1893 World's Columbian Exposition in Chicago, all temporary except the Parthenon, which was kept and rebuilt years later to cement Nashville's self-image as "the Athens of the South." Nashville, with several prestigious schools, promoted itself as a city of learning like the ancient Greek capital. It was a capital besides, a reasonable claim on charging ahead as a major Southern power center.

NBW considered the exposition's embrace of past and future as it sought to strengthen the park's distinct spatial lineages, both formal and less so. The firm looked for ways the region's ecologies had shaped its cultural history, a search that revealed pivotal but obscured past developments. Certain historical findings later supported the placement or relocation of commemorative elements along an enhanced system of walking trails through the outer groves, to which the firm would add ornamental trees among the dense hardwood canopy.

Middle Tennessee's geology helped explain its ecological profile. Geologists describe the region's limestone karst substrate as the remnant of an ancient sea. Limestone and shale formed a cap that tectonic forces heaved into what geologists call the Nashville Dome. The karst's watery cavities well up in numerous springs, and the shale's erosion exposed salt licks that drew bison, elk, and other herds along pathways called buffalo traces. The pathways became continental passages to Euro-American explorers and settlers. One path, the 500-mile Natchez Trace, runs from the lower Mississippi River to its northern head at Cockrill Spring inside the southwest edge of Centennial Park.

Around 1780, the land around Cockrill Spring was farmed by the family of Ann Robertson Cockrill, said to be the first woman to gain title to land west of the Appalachians. Travelers on the Natchez Trace knew of the spring's excellent water. Among the travelers were itinerant singers and fiddlers who formed artistic communities—the beginnings of Nashville's renown today as the seat of roots-music royalty (a society the director Robert Altman

captured in his 1975 film *Nashville*, which culminates at the Parthenon).

Cockrill Spring was buried in the mid-1800s to prevent the waterborne spread of cholera after repeat epidemics. NBW was first shown Cockrill Spring under a utility cap. In 2014, Nashville's mayor removed the cap to launch the spring's daylighting by NBW. Today the spring flows up to a limestone terrace, runs through a constructed wetland glade, and enters Lake Watauga across the park to irrigate the grounds. Through the glade, NBW laid two low limestone walls, like old foundations, in tribute to the Natchez Trace. A stone marker nearby commemorates Ann Cockrill.

In the same quadrant, NBW's new amphitheater, Musicians Corner, salutes traditional musical forms that germinated around Nashville and the Natchez Trace and evolved into pillars of global entertainment. The earthen performance space (home to a Musicians Corner festival) backs into a modest berm that evokes ancient Mississippian mounds behind benches stretched in broken rings, an ensemble reminiscent of camp settings along wilderness trails.

Centennial Park was the site of a crucial victory for the women's suffrage movement. A bronze likeness of five leading suffragists, by the Nashville sculptor Alan LeQuire, was placed on a walkway toward the Parthenon in 2016 to commemorate a 1920 rally to press Tennessee's legislature to ratify the 19th Amendment. The amendment's narrow passage made Tennessee the thirty-sixth and decisive state to approve the constitutional right of women to vote in the United States. The suffragists' figures appear to stride toward the Parthenon's steps, larger enough than life to convey the magnitude of their triumph.

Since 1996, a Children's Memory Garden has been a space for remembering local children killed during acts of violence. NBW reconceived the garden as intimate terraces between curving pink-granite walls with seating to invite meditation. Small LED ground lights pulse gently after dusk. The lights are the work of Tillett Lighting Design Associates, the designer of the Parthenon's new lighting. At twilight, the building illuminates in stages: first its bronze doors, then the inner walls—a soft white glow or, on occasion, specified colors—then the entablature, the tympanum, and finally the griffins atop the corners.

Views of the Parthenon's corners in Athens figured in calibrating approaches to it on the Acropolis. NBW studied the early Greek arrangements in its planning of path geometries around Nashville's exact replica of the temple. At a more magnified scale, the design revived a Beaux-Arts formality to frame the Parthenon and the Great Lawn, drawing allées of vigorous trees along the lawn's edges, with wide paved walkways for staging outdoor fairs and carrying service vehicles, studded by low bays of rusticated stone that provide visual rhythm, tree protection, and seating. The walkways stretch long toward the Parthenon and build a fitting sense of awe as the body draws nearer and the mind appraises its tremendous weight and its significance for Nashville and the nation that Tennessee joined as the sixteenth state.

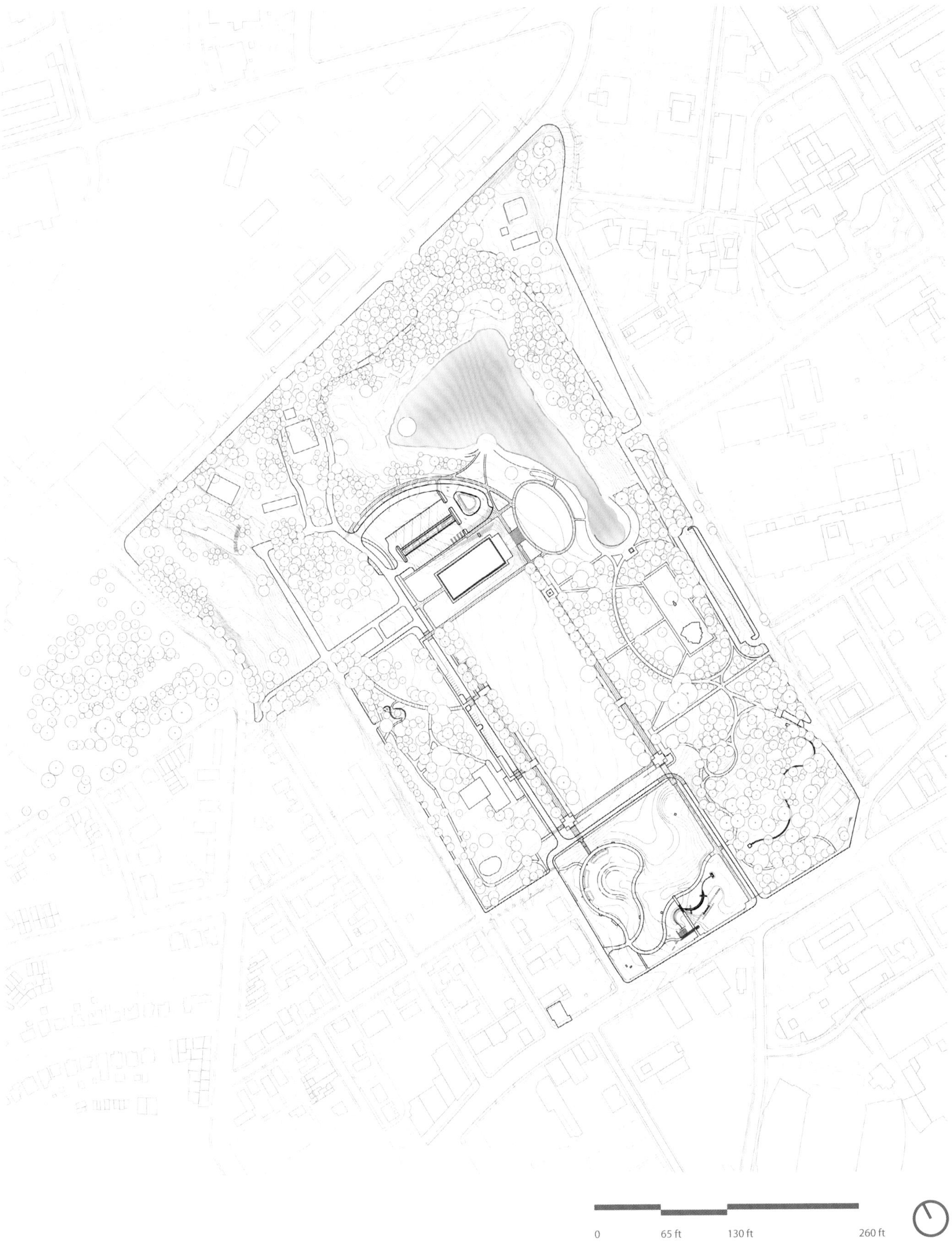

↖ ↑ In the NBW plan for Centennial Park, the geometry of core areas around the Great Lawn and Parthenon are drawn from Beaux-Arts traditions of strong, clear axes to reinforce the heroic scale of large, bright spaces and long views. Around the core, Picturesque paths meander among the many large trees and around Lake Watauga for more intimate experiences in the relief of ample shade.

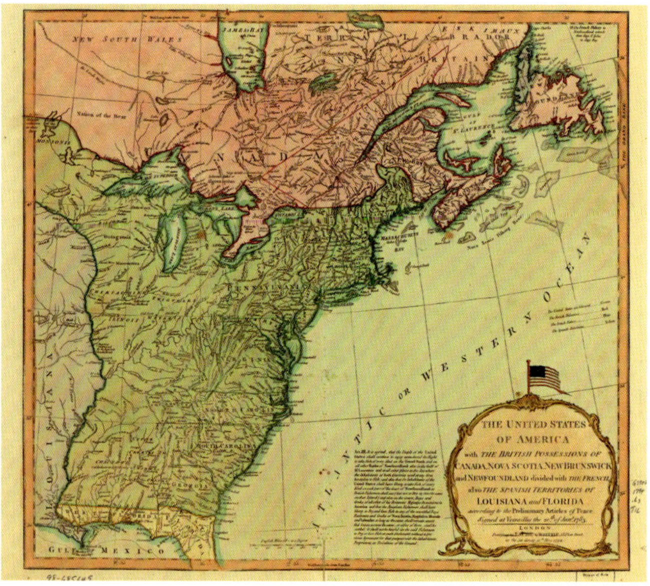

Map of the United States two years before Tennessee statehood, Laurie & Whittle, 1794.

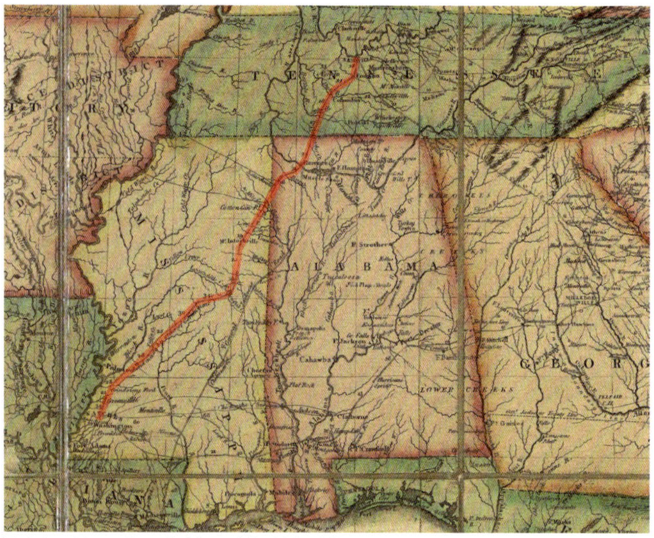

Natchez Trace (in red), Map of the United States of America, J. Melish, 1820.

Bison drawn to Nashville's Great Salt Lick shaped paths later followed by hunters, explorers, and settlers.

Bird's-eye view of the Tennessee Centennial Exposition, Henderson Lithographing Co., 1896.

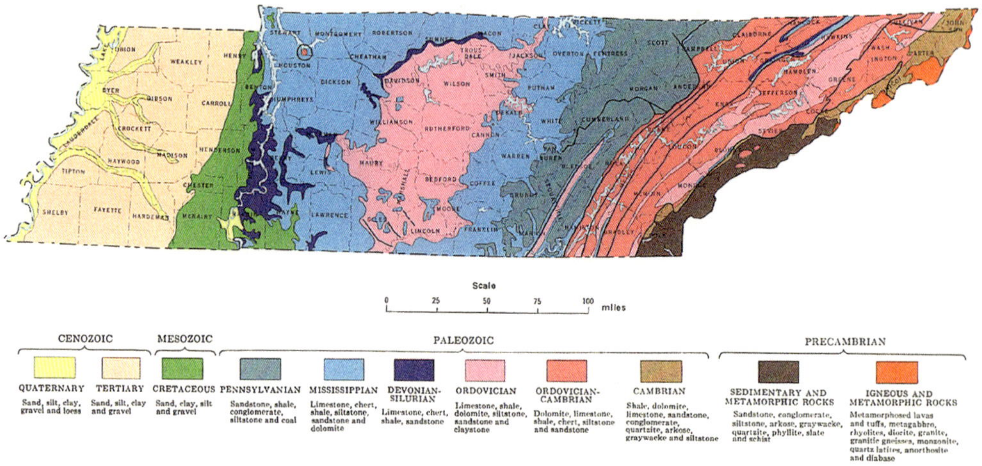

Geology of Tennessee. Ordovician limestone and shale of Nashville in upper left of pink region.

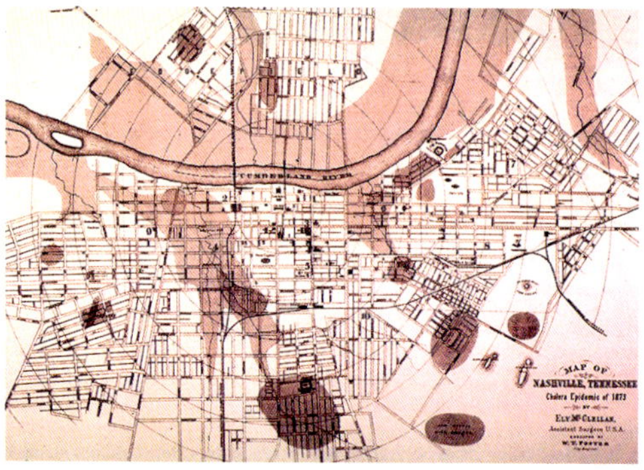

Repeated cholera outbreaks, like the epidemic of 1873 mapped here, led the city to divert streams below ground to achieve safer drinking water.

Gustav Fischer's Centennial Band, resident entertainers at the 1897 Exposition.

Pharr Mounds along the Natchez Trace in Mississippi, dated c. 1–200 CE.

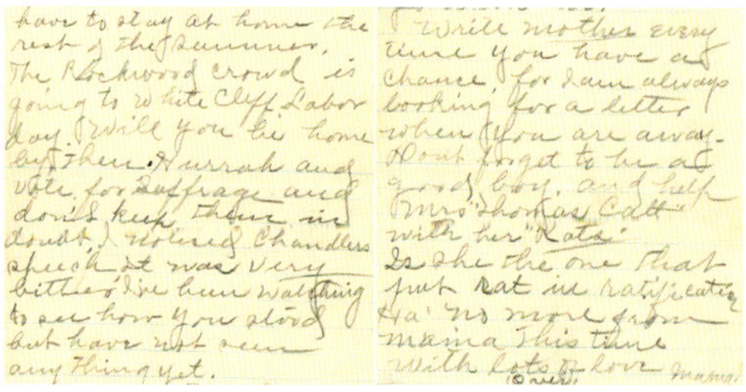

have to stay at home the rest of the summer. The Rockwood crowd is going to White Cliff Labor Day. Will you be home by then. Hurrah and vote for Suffrage and don't keep them in doubt. I noticed Chandlers speech it was very bitter. I've been watching to see how you stood but have not seen anything yet.

Write mother every time you have a chance, for I am always looking for a letter when you are away. Don't forget to be a good boy, and help Mrs. Thomas Catt with her "Rats." Is she the one that put rat in ratification, Ha! No more from mama this time

With lots of love Mama

Suffragist Febb Ensminger Burn's letter to her son, Congressman Harry T. Burn, urging him to vote to ratify the 19th Amendment.

↑ NBW's strategy revives the long-buried Cockrill Spring, which emerges from the region's karst geology. The spring was known to people arriving in early Nashville on the Natchez Trace, which is represented by broken lines of limestone. The spring water resurfaces within a limestone terrace.

↑↑ The spare, rustic Musicians Corner celebrates the musical culture that arose organically in Nashville around the Natchez Trace. ↑ Cockrill Spring's water courses along a sinuous channel through a wetland glade and is recaptured in Lake Watauga for use in irrigating the park. → Musicians Corner extends the Southern music traditions it represents in an annual series of performances.

MUSICIANS CORNER
Coleman

↖← The Great Lawn stretching south from the Parthenon is a heavily socialized city green for leisure, recreation, and large public events. ↑ A bronze sculptural group by Alan LeQuire of five leading suffragists stands on an approach to the Parthenon, where in 1920 the women successfully rallied for passage of the 19th Amendment.

↑→ NBW expanded the Children's Memory Garden west of the Parthenon. It is meant as a protective space, with stone seating that emerges from the ground as a gesture of embrace, a young Willow Oak tree placed near an existing mature Willow Oak to signify guardianship, and at dusk, an array of small, softly pulsing lights embedded in the pavement.

NATURE FOR EVERYONE

PROJECT
LOCATION

BOK TOWER GARDENS
LAKE WALES, FLORIDA

As the editor for thirty years until 1919 of *Ladies' Home Journal*, Edward W. Bok strived to improve the lives of Americans by urging comfort and balance in their daily routines. He is said to have first popularized the words "living room" from his perch. His own primary living room was outside Philadelphia, but in winters he headed for his house in Lake Wales, at the center of the Florida peninsula. He was captivated by the region's landscapes and especially by a rise of land nearby called Iron Mountain. He bought 250 acres of this highland in 1922 and began to create the public gardens called Mountain Lake Sanctuary and Singing Tower, now known as Bok Tower Gardens. He had the gardens designed by Frederick Law Olmsted Jr. and commissioned an opulent stone carillon by the architect Martin B. Medary.

The sanctuary was his gift to Americans. Bok was a Dutch immigrant who felt indebted to the United States for the eventful life it opened to him. President Calvin Coolidge dedicated the sanctuary in 1929; tens of thousands of people attended—or tried. One signal of Bok's broad inclusiveness was to have parking for no fewer than five hundred cars. He conceived the sanctuary as a show of noblesse oblige to the nation, "a place of quiet and repose for the electrically driven people of America." He saw Americans as too removed from art, nature, and beauty and wanted the gardens to "express the gospel of beauty: to open our eyes and awaken our senses to the beautiful." The carillon, descended from a beloved Dutch tradition of bells cast in bronze and hung high over towns, added a distinctive sonic experience to the garden's sights and fragrance.

The gardens were constructed as a composite of lush Floridian subtropics carried to Iron Mountain's austere sandhill summit. The mountain, part of a narrow spine in central Florida, rises about three hundred feet above sea level, one of Florida's highest points. Bok's generosity toward people extended as ardently to wildlife. With Olmsted, he plotted to attract the maximum number of birds, native and migratory, to the sanctuary by installing a variety of plants that would supply them with berries and other food.

NBW began work at Bok Tower Gardens to build on a 2011 master plan by Studio Outside of Dallas commissioned by the directors as a long-range strategy for the garden as it neared its centenary. NBW's research and design focused on the evolving ecological needs of the landscapes in an age of climate chaos and imperiled biodiversity. The grounds had changed in aspect over the decades as plantings thickened and shade increased, but the property retained much of the original layout of paths, plantings, and open areas. The NBW design also was aimed at making the spaces easily accessible to people on foot or with walkers, strollers, or wheelchairs—and easy to navigate, from the launch point of a large oval lawn near the visitor hub, by following discreet signage mounted on low walls.

The firm sought to amplify the original vision for the garden as an intensely productive environment where nature and art edify and intrigue people—children not least—as they wander, discover, and lose themselves among its winding paths. The run-up to design involved immersion in Bok's and the designers' correspondence, design documents, and notes. It included a review of visitation analyses alongside likely trends among visitors. With the garden administrators and staff,

the NBW team walked the site at various hours to observe activity among people and animals and to note the qualities of light and air. They assessed the ways alterations of trees or paths would affect progressions and views.

It was crucial also to estimate the impact of renewing the ecologies of the garden, or introducing appropriate novel ecologies, on its existing natural systems. At the outset, Bok and his designers had taken pains to amend the site's substrate of Candler Fine Sand with thousands of truckloads of rich black dirt to ensure the longevity of the original investment in plants. They estimated losses of around 20 percent during an acclimation period, though just 5 percent of the initial plantings were lost.

In this recent round, NBW consulted with professionals in the region—ecologists, designers, horticulturalists, growers, and restoration specialists—and studied nearby nature preserves to steer toward solutions aligning with the emerging dynamics in the region's environments. The design anticipates intensifying rainfall, reasonable to expect as the warming atmosphere carries more moisture. The site's sandy ground accepts large amounts of rain. Excess loads from the outer ridges and the parking lot now flow down to a new central infiltration zone in a valley and collect in a stormwater basin inside the Discovery Zone.

Nearby is a prodigious food garden, where children can learn from interpretive panels about aquifers and protecting water quality amid many other connections to make in this living library of central Florida. The addition of plainspoken interpretation elements promotes a passion for advocacy in the spirit of Edward Bok. People of any age can engage with the subtle annotations around pollinator plantings, endangered Florida plants, the critical role of fire, and the "domino effect" of extinction that imperils, for one, the gopher tortoise that inhabits the scrubland of the sandhill environment along the site's eastern side.

The state of Florida has listed the gopher tortoise as a threatened species. It is also what is known as a keystone species for the ways it supports other species and the ecological health of its specific habitat. Its burrows serve as shelter to three hundred other animals, and its feeding on wiregrass and other plants sustains the sandhill's open character. The Discovery Zone informs kids in multiple ways about the gopher tortoise's outsized importance to its surroundings, crucial connections to sow among young people, given that the species will survive or disappear depending on a new human awareness and action.

↖ ↑ The original Olmsted pathway to the bell tower had been lost. NBW planned the circulation through the expanded gardens to bring visitors to Olmsted's original tower approach and offer the experience of the authentic vision. The Ellipse holds a central lawn and serves as an orientation device, opening marked thresholds to multiple paths at its edges.

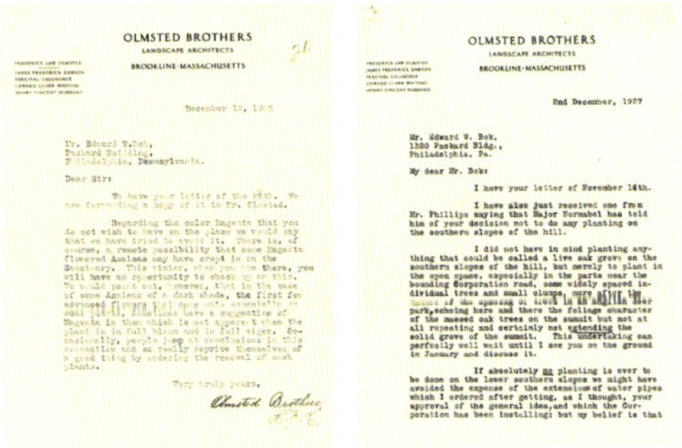

OLMSTED BROTHERS
LANDSCAPE ARCHITECTS
BROOKLINE-MASSACHUSETTS

December 18, 1925

Mr. Edward W. Bok,
Packard Building,
Philadelphia, Pennsylvania.

Dear Sir:

We have your letter of the [illegible]th. We are forwarding a copy of it to Mr. Olmsted.

Regarding the color Magenta that you do not wish to have on the place we would say that we have tried to avoid it. There is, of course, a remote possibility that some Magenta flowered Azaleas may have crept in on the Sanctuary. This winter, when you are there, you will have an opportunity to check up on this. We would point out, however, that in the case of some Azaleas of a dark shade, the first few advanced flowers [illegible] have a suggestion of Magenta in them which is not apparent when the plant is in full bloom and in full vigor. Occasionally, people jump at conclusions in this connection and so really deprive themselves of a good thing by ordering the removal of such plants.

Very truly yours,

Olmsted Brothers

OLMSTED BROTHERS
LANDSCAPE ARCHITECTS
BROOKLINE-MASSACHUSETTS

2nd December, 1927

Mr. Edward W. Bok,
1320 Packard Bldg.,
Philadelphia, Pa.

My dear Mr. Bok:

I have your letter of November 14th.

I have also just received one from Mr. Phillips saying that Major Hornabel has told him of your decision not to do any planting on the southern slopes of the hill.

I did not have in mind planting anything that could be called a live oak grove on the southern slopes of the hill, but merely to plant in the open space, especially in the parts near the bounding Corporation road, some widely spaced individual trees and small clumps, more after the [illegible] of the spacing of trees in an English deer park, echoing here and there the foliage character of the massed oak trees on the summit but not at all repeating and certainly not extending the solid grove of the summit. This undertaking can perfectly well wait until I see you on the ground in January and discuss it.

If absolutely no planting is ever to be done on the lower southern slopes we might have avoided the expense of the extension of water pipes which I ordered after getting, as I thought, your approval of the general idea, and which the Corporation has been installing; but my belief is that

Project correspondence from Olmsted Brothers to Edward Bok, 1925 and 1927.

Maquette of Singing Tower, Martin B. Medary, architect, 1927.

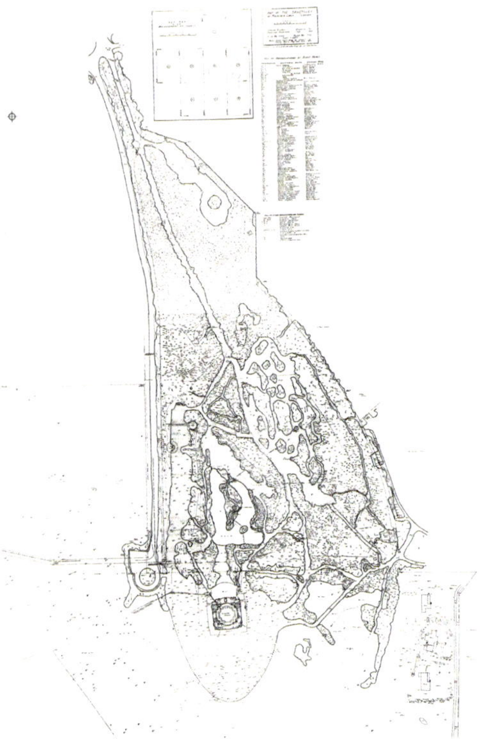

Map of the Sanctuary of Mountain Lake (detail), Olmsted Brothers, July 1930.

Singing Tower during construction in 1927.

Carillon, designed and built by John Taylor Bellfoundry, Ltd., Loughborough, England.

Installation of Singing Tower carillon bells, 1928.

Edward Bok intended to create a sanctuary for local Floridian and migrating birds, including the Roseate Spoonbill and the Great White Heron, both illustrated in *Audubon's Birds of America*.

Gopher tortoise, a keystone species in the sandhill environment.

Prescribed burning, beneficial to the threatened gopher tortoise.

↑ The project directed considerable attention to restoring the Florida sandhill ecology around the central gardens; prescribed burns discourage insurgent species that otherwise displace the open canopy of longleaf pines and turkey oak and the understory of wiregrass, the critical habitat of the gopher tortoise. →→ Near the Discovery Garden, a trail leads visitors from the formal areas of the grounds to areas conserved as sandhill habitat.

← Olmsted and Bok scripted the historic core's winding, tree-sheltered walk to bring visitors to a pond that mirrors the tower and its frame of moss-draped hardwoods and palmettos. ↑↑ A path bends along an overlook westward where views stretch fifty miles. ↑ An interior node orients visitors between two main garden paths. →→ In the Florida Garden, the bog boardwalk leads through a large clearing between sections of cloistered paths.

↑↑ The pollinator garden introduces visitors to the role of birds and insects in Florida's natural food web. ↑ Paths wind at various scales in circles and spirals among the collections. ↗ Hard surfaces on the restored Olmsted approach to the tower allow universal access. → A boardwalk floats through the bog garden, with seating to encourage extended observation by visitors.

↑ → The extensive Discovery Garden offers all visitors, especially children, an empathetic sense of the region's animal species, as in the panther habitat of simulated limestone karst springs, a giant bird's nest, a gopher tortoise burrow, and pollinator plantings. These settings inspire children to "become" the animals and interpret their ecological importance through play. →→ Near the center of the Discovery Garden is a dry karst pond, or sink, for adventure play, with small gathering areas at its edges.

↑ The kitchen garden emphasizes public awareness of sustainable agriculture and health in the growing of fruits, vegetables, and herbs and in cooking demonstrations held in the outdoor kitchen. → The open lawn of the Ellipse provides space for concerts and other large public events.

THE SECRET BELOW THE SURFACE

PROJECT — HUDSON YARDS PUBLIC SQUARE AND GARDEN
LOCATION — NEW YORK, NEW YORK

NBW served as the lead landscape architect for the Hudson Yards megaproject, which broke ground in 2014 and has added 12 million combined square feet of residential, commercial, and office space so far to a 28-acre area between West 30th Street and West 41st Street near the Hudson River edge of Manhattan. The development also includes the Shed, a performing arts space created beneath a shell that moves on wheels to serve as a roof, and the Vessel, a monumental sculpture of connecting spaces conceived as a viewing platform. The land in this instance of landscape was notional; Hudson Yards is on entirely invented ground, a harvesting of air rights made possible by constructing a technically complex platform over the live railroad tracks that serve Pennsylvania Station a few blocks east.

Above the Public Square, the towers of Hudson Yards are visible for miles, as if a new mountain range had sprung from the West Side of Manhattan. In parts of the six acres of public spaces at their base, it's easy to imagine the towers aren't there. Around the square's wide-open expanses are banks of huge trees with shrubs underplanted in high, dense drifts that create enclosure and cooling shade. One long edge has loose seating and tables under a bosque of black gums in full pirouette to form a solid canopy, and trees rise as high as 50 feet from the promenades, with heavy crowns that sway in breezes against backdrops of glass and blue sky.

The plaza and its plantings confront the elements head-on, given a location that faces west toward the Hudson River. With periodic certainty, the sky darkens, the breeze becomes a gale, and there is unrelenting summer heat. The microclimate came under consideration early as the firm began composing the plant communities of Hudson Yards. The planting scheme is conceived as a pocket of nature farther north in the Hudson River Valley—forest depths, edges, and openings—though Hudson Yards gets little reprieve from its harsh exposure to sun, wind, cold, warmth, rain, snow, ice, and, not least, from the handicaps imposed by the colossal buildings and infrastructure that encase the site—more heat, higher wind, and limited room for planting. To make the investment in a living landscape viable, it needed insurance, which mainly takes the form of soil.

Because it is an air-rights development, there was no soil to start—all soils would be brought in to fill the platform once it was built above the working tracks of a storage yard for the Long Island Rail Road and, below them, the tunnels that take Amtrak and New Jersey Transit trains under the Hudson River. The soils for the project were specified to set up the horticulture with its best chance to thrive. Given the constrained root zones below the surface, the soil consultant designed a structural soil that goes long on sand, augmented by compost, nutrients, mulch, and microorganisms to establish a symbiotic energy and food web around the roots. This high sand quotient results in a forgiving density to let the roots spread outward given that they will hit the bottom of the planting volume early in their development.

The roots also must be protected from the heat coming from the trains below the platform, which can reach 150 degrees Fahrenheit. A system of tubing runs chilled water and glycol throughout the root zones. Sensors track temperatures throughout the soil at a six-inch depth. A reading of 75 degrees activates the coolant to turn

temperatures down to 70 degrees. The landscape and buildings also hold powerful horizontal fans to pull heat out of the track areas beneath the square's platform.

All the calculation backstage, or below stage, is to support a set of spaces that are broadly sociable. There are spaces ample for festivals or for crowds to watch baseball on the big screen and others suited to smaller gatherings or for those who simply come to read, have a drink, or watch people. Seating is in high supply along hundreds of feet of site walls in the main plaza, scaled to leaning or perching. At an outer edge of the development, at Tenth Avenue and West 30th Street, a group of black granite benches surrounds a stair up to a spur of the High Line, accompanied by river birches planted in rain gardens and a low stone fountain carved to represent a geologic extrusion of the Hudson River Valley, a human-scaled ensemble to temper the muscular steel supports for the heavy platform overhead.

Connection and movement define the plaza's constructed forms. Its design makes seamless transitions to the High Line's reaches southward and, to the north, Hudson Park and a new subway entrance for the 7 line. The main walkways carve arcs through the center that anticipate the natural movements of people through this quarter with intuition an orthogonal pattern would stifle. Amid the mottled gray paving are long arcs of black pavers that mark the memory of the early train tunnel into Penn Station beneath the site—a detail that emerged during the firm's site research—and have the effect of drawing the body gently forward. The spatial experiences throughout the Public Square, plenary or personal, gain energy from the varied shifts in grade among spaces that connect by gentle slopes or stairs that flare from narrow to wide in their descent, and with stone-lined edges that melt from honed to rusticated. At ground level, beneath the masses of the towers, the emotional temperature of the plaza is calmed by a palette of stone pavers in subtle keys of gray, a surface that conceals the complex support system for the landscape encased beneath it.

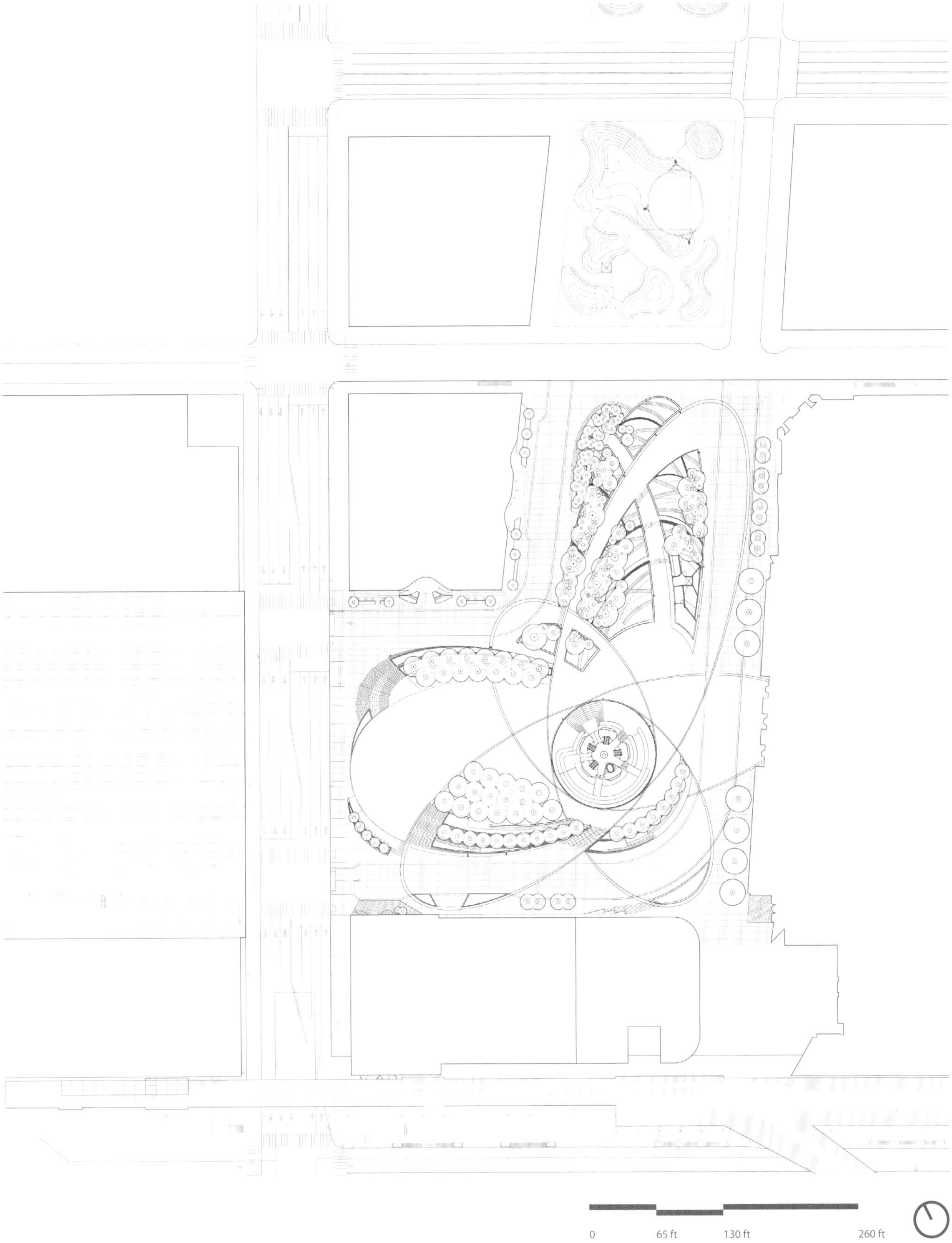

↖ ↑ The Public Square spreads out across an intensively technical platform constructed above twenty-eight working rail lines. On its surface, interlacing elliptical lines in the paving, centered on the Vessel, create visual circuits amid the towers and connect people to surrounding neighborhoods, avenues, streets, gardens, terraces, and the adjacent High Line.

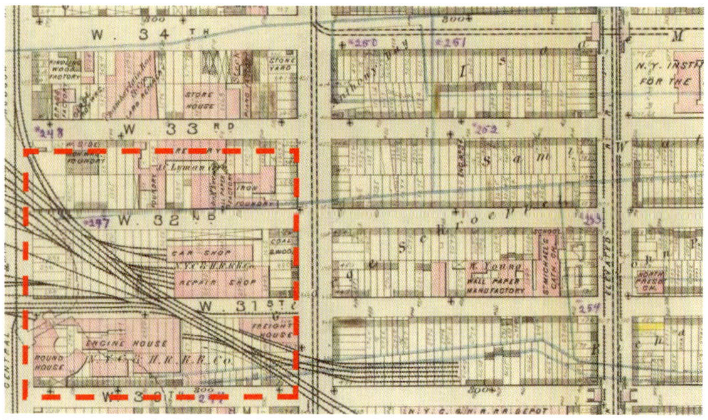

Map annotated with Hudson Yards extents, Part of Ward 20, New York City, shows West Side rail complex, 1879.

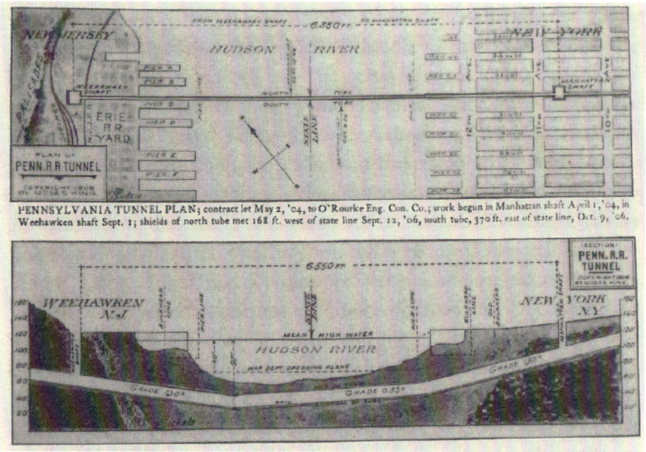

Plan and section of Pennsylvania Tunnels under Hudson River, 1906.

Excavation workers with tunneling shield, 1905.

An early concept of the Public Square featuring a structure evoking twentieth-century engineering on the Hudson Yards site.

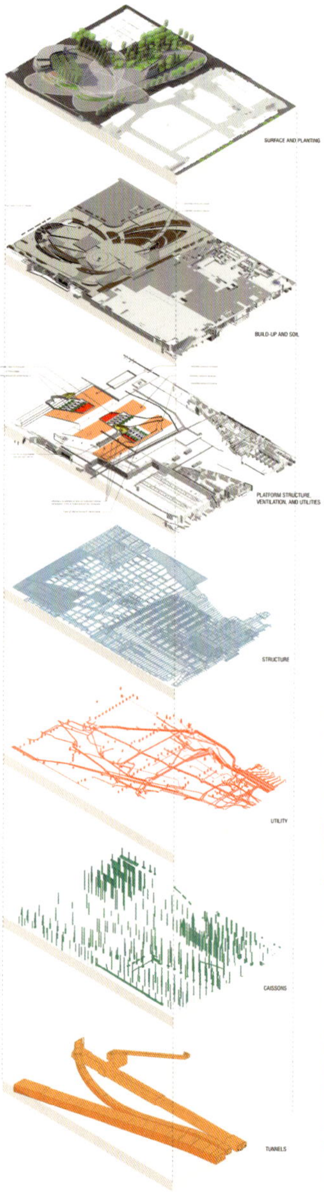

Public Square platform, exploded layer diagram of plantings, soils, structure, utilities, and rail lines.

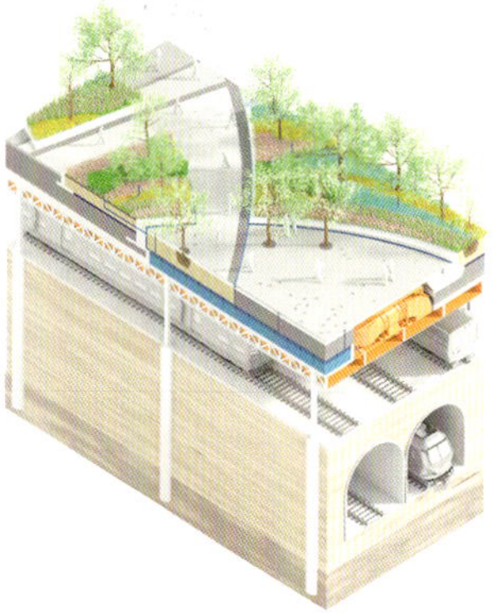

Public Square platform, subsurface infrastructure.

Mannahatta: A Natural History of New York City was referenced to select the native-tree canopy that historically occupied this area of the city.

Highly technical soil beds were designed to provide the space, irrigation, and cooling necessary for plantings to survive.

To provide mature trees on opening day, the trees acclimated in a nearby nursery for three years before several all-night installation efforts.

↑↑ At the south edge of the Public Square, a bosque of native black gum trees shelters an informal seating area.
↑ Broad stairs bounded by granite seat walls connect garden terraces.

↑ The Public Square occupies a pocket position among towers that required an analysis of sun and shade patterns; trees were nursery-grown in simulated conditions to acclimatize them after installation. ↗ → Stratified walls and stone benches make reference to the underlying granite geology of the site. →→ Broad walkways carve long arcs across the plaza and among terraces, similar to the desire lines people wear into landscapes along preferred routes.

↑↑ A stone fountain marks the corner of Hudson Yards beneath vertical infrastructure at Tenth Avenue and West 30th Street. ↑ A fissure in the fountain's form represents the broken rock, or breccia, found between adjacent plates in the Hudson River Valley.

↑ →→ Black gum trees on the north edge of the Public Square form a shaded loggia of trunks along a bench that seats dozens of people. Each wood member of the bench is slightly different from the others; from end to end, the ratio of back height to angle shifts from a modest recline to an upright posture.

↑ Near Tenth Avenue and West 30th Street, rain gardens hold river birches among a series of stone benches.
↗ → The paths, with variegated stone paving, are wide enough to hold large numbers of people yet shrouded by sizable trees to create intimacy and the sense of the Hudson Valley forest environment.

II

ESSAY 76 LANDSCAPES OF CARE: DESIGNING FOR AN ECOLOGY OF EMPATHY
Nina-Marie Lister

PROJECTS 84 BROOKLYN NAVAL CEMETERY LANDSCAPE
Brooklyn, New York

96 DUKE UNIVERSITY RECLAMATION POND
Durham, North Carolina

108 THE RAMBLE
Fayetteville, Arkansas

ESSAY

LANDSCAPES OF CARE: DESIGNING FOR AN ECOLOGY OF EMPATHY

Nina-Marie Lister

petal soft
  certain pages
    of her field guide
      —Laurie Greer, 2018[1]

The only way to learn a landscape is to soak in it. Walk, stand, and sit with its silence and symphony, be present in the place of its making, keeping, and becoming. Look, listen, feel—steep in its sensorium. Breathe in the spectrum of its scents, bathe in its soundscape, attend to the theater of its light and shadows, and be wrapped in the tapestry of its textures. This is not fast work. It is a slow awakening to growth that cultivates awareness, understanding, appreciation, and perhaps, revelation. This is the place from which hope is sown and grown.

We desperately need hope now, as we face down a global polycrisis, the intertwined and cascading crises of climate change (global heating), biodiversity loss (mass extinction), and socio-ecological vulnerability (human suffering). These entangled crises are at once cumulative and synergistic. Without urgent intervention, there will be dark times of our own making, not only for humanity but for the diversity of other species with which we share the planet. But nature, and all life on earth, is not merely collateral damage in humanity's downturn, it is the key to our own survival. The agency of landscape architecture in facing this integrated and complex challenge is clear. Beyond the apparent prestige of STEM status recently bestowed on the discipline,[2] the material of making landscape is, at its core, alive. It is *ecological*, as we too are ecological beings, dependent on and integrated with the abiotic, biotic, and cultural processes of life on earth—a relationship with the *in-between*.

As a field ecologist, I learned the quiet patience of distance and the perseverance of systematic observation. Later, as a designer, I un-learned the dominance of left-brain training to accept and embrace the freedom of stepping out of bounds, and to appreciate the role of intuition in creativity. I plunged into the in-between, belaying its inscrutable depths. This is where the hard work of interdisciplinarity lies, and from it, the capacity to integrate science and art. As I will argue here, the landscapes of NBW master the in-between; they reveal an embedded process that celebrates an epistemology of *both/and*, skillfully weaving and joyfully wandering across the intersection of science and art.

This is poignantly evident in two projects: the pocket meadow of the Brooklyn Naval Cemetery Landscape and the expansive prairie and land bridge of Memorial Park in Houston. These are two distinct landscapes created with equal care, both underscored by an ecology of empathy. Counterpoints across two scales, these landscapes result in places of revelation and hope.

Meandering in a Meadow: Brooklyn Naval Cemetery Landscape (1)

fig. 1 Author's photographs of the boardwalk and the bench.

As I walk along Williamsburg Avenue, the cemetery landscape entrance rises gently from the sidewalk, a slow and dignified uplift along a quiet wall that leads to an elegant gate and visitor pavilion. There is an intimacy of scale, as I pass through the gate, the sense of a secret garden revealed. The landscape opens into a wide embrace, in a wash of waving grasses, heads heavy with September's seeds, a backdrop to purple asters and goldenrod, twin-blooms exuberant in late summer's glow. I am at once held by the hum of insects and struck by the silence of the city.

At 1.7 acres Brooklyn's Naval Cemetery Landscape is a tiny oasis in an urban post-industrial corner. Prior to Dutch-colonial settlement, the site was more marsh than land, formed by a creek that flowed through a wet meadow to the river. Centuries later, the site was home to the Brooklyn Naval Yard's former hospital, and the cemetery lay behind it. These layers of history, from pre-settlement to colonial to post-industrial, are present in the park's understated and immersive design, as much as they are integral to the process that realized it.

The story of the cemetery-park is compelling, and not a little haunting. Over a century from 1700 through the late nineteenth and early twentieth century, more than 3,000 human bodies from the hospital and naval yards were buried in unmarked graves on the site. Most human remains were exhumed and removed before the project began, but the unmarked nature of the cemetery leaves uncertainty, and inevitably, some remains. For reasons of heritage preservation and cultural respect, the design team was permitted to work only in the top four inches of soil, and anything below was to remain undisturbed. Closure is of an emergent nature here: the design team used this constraint as an opportunity to encircle the burying ground with a raised boardwalk, thereby making sacred the space within. Granite blocks recall mooring posts for early naval ships, linking the meadow to the memorial grove—like stepping stones for spirits and clambering children alike.

Quite literally, the project was an exercise in land*scraping* to prepare the surface of the ground for renewal. Selected from regionally specific sources, the seeds of native meadow plants were pressed into the top layer of soil, from which grasses and seasonal flowering perennials have emerged. The wet meadow is now a colorful palette of pollinators, restoring life to honor the dead. This open meadow is encircled by an understory of shrubs against the backdrop of newer young and existing mature trees. Local community groups steward and care for the landscape, leading learning tours and outdoor classes. Their care and commitment are quietly present, evident as a yellow-vested worker crouches among the plants, removing invasive species and maintaining the meadow. Pamphlets on pollinator identification and butterfly watching invite visitors to explore, and a meditation journal reveals reflection and respite across the ages.

Seven years on, the flora have burst free of their planting design, promiscuously wandering across the site, as do the foraging birds, bees,

fig. 2 Author's photographs of the wetlands boardwalk and blocks of up-cycled roadway in Memorial Park.

butterflies and moths, a joyful cacophony of abundance, and a testament to life as it renews a place of death. The boardwalk appears to hover, a sinewy "river of wood in a sea of grass," as Woltz aptly describes it. Meandering around the meadow, the boardwalk opens to an amphitheater under the shade of mature red maples and a few evergreen trees, and continues around to a quiet grove, with a bench for contemplation.

Walking the "wooden river" amid a flourishing meadow is at once an immersive and meditative experience. My undisciplined mind wanders too, recalling *solvitur ambulando* (it is solved by walking), as the maxim of walking meditation, an exercise undertaken in many parks worldwide. Meditative walking, sitting, and reflecting are encouraged here, on a healing journey that is itself, central to the design intention of the cemetery landscape. The project was funded in part by the not-for-profit Nature Sacred[3] and its founders at the TKF Foundation, an organization dedicated to healing spaces. The group invests in sites of biophilic healing potential and gathers data on the human health and wellness benefits of nature and biodiversity. Meditation journals remain on-site and are filled by visitors who share their thoughts and impressions, later collected as part of an ongoing psycho-social and environmental research project. The cemetery landscape is an integrated living experiment, a laboratory for learning and healing.

Equal parts contemplative space and performative project, the naval cemetery landscape is one of understated simplicity and quiet complexity. Biodiversity bursts onto (and out of) center stage in a meadow of performance art, a theater of pollination in a field of forgotten sorrows. Life continues, on a new path, emergent and hopeful. I feel it here in this small urban meadow, sitting with my field guide, my friend Thomas Woltz. We rest in silence, in this bowl of sun, faces turned to its low autumn glow, ears tuned to the thrum of ground nesting bees, playing bass to the whispery-whistle of fall warblers foraging in the dogwoods. The day is light and heavy, as I grieve the loss of a dear friend who left us that morning.[4] I blink hard into Wendell Berry's "day-blind stars,"[5] and a world of wonder is revealed in this pocket of a Brooklyn corner.

The cloistered cemetery landscape proves a powerful counterpoint to the expansive Houston Memorial Park, where wet and dry prairies meet, bridged by a new landscape altogether.

Bridging the Prairie: Memorial Park 2

I had the gift of an early spring, walking through Houston's newly expanded and reconnected Memorial Park on opening day in February. It was chilly (for Houston), wind whipping the saplings in their freshly made beds, bending thousands of tiny plugs of native grasses, sedges, and reeds flat to the new soil. Opening day is always underwhelming for the landscape architect. Unlike architects, whose new buildings show perfect and pristine, no fingerprints or scuffmarks yet to be seen, the landscape architect must conjure the future beyond bare soil and skinny saplings: imagine all this when the trees are mature, the shrubs grown up, and the beds filled in! But Houston didn't care. No small affair, opening day and the "biggest picnic in Texas"[6] drew crowds of Houstonians in obvious celebration of their beloved, renewed park.

Houston's century-old 1,500-acre Memorial Park was once an urban wilderness. Starting in the 1940s and 1950s, it was ruptured by a burgeoning Memorial Drive, eventually expanding into a high-speed multi-lane road that split the park in two and divided its communities, human and wildlife alike. In this portion of a larger rethinking of Memorial Park, the landscape has been re-integrated and expanded through NBW's addition of the Kinder Land Bridge and the Cyvia and Melvyn Wolff Prairie. These two new landscapes functionally reconnect the north and south sections of the park, which had been severed into isolated fragments by the drive. In addition to ecological restoration, this work is principally a living landscape infrastructure designed to provide improved connectivity, stormwater infiltration, and flood management as well as reintroduction of native biodiversity in restored habitats. The land bridge itself is a significant structural and transportation engineering undertaking, less a bridge than a tunnel for cars with a living roof, a new landscape of fifty-some acres that spans four lanes and reconnects the park.

Adjoining the land bridge, the prairie is a fifty-five-acre complex ecological design project. Much of the park's tree canopy was in poor health, warranting removal of diseased and dying trees while healthy mature native trees of high socio-ecological value such as loblolly pines, Live Oaks, and Red Oaks were relocated to the restored park section. Endangered, native Gulf Coast prairie habitat has been reintroduced and restored, along with a wet meadow and an ephemeral creek, prairie potholes, a dry prairie grassland, and a savannah. An undulating boardwalk loop opens views across the wetlands and meadows to the trees beyond.

A series of elegant follies provide shelter, shade, and quiet respite, their slatted sides split the sunlight, scattering light and shadow, patterns that dance with the sun. Even in the cold light of early spring, the landscape shimmers, the clear creek a mirror to the sky, grasses rustling and Live Oaks sprawling. Through it all, a winding series of new trails and a grassy common grace the top of the land bridge, accessed too by an Anthropocene scramble—created by the reuse of the old Memorial Drive material excavated in the process—to new trails and the boardwalk to reconnect and extend the existing trail network from the north and south park sections. Together, the land bridge and new prairie complex serve as a powerful healing design, a lush living layer over an old wound.

The prairie has long been a storyline through, if not a central figure in, the work of NBW. The firm's early projects focused on restoring agricultural and ecological productivity to farm sites through the establishment of grassland meadows, savannahs, and prairies. This expertise is captured and reflected in a diversity of conservation-restoration projects in California, Virginia, and New Zealand. In the firm's first monograph, *Garden Park Community Farm*, design critic

fig. 3 Memorial Park visitors, photographed by the author.

Beth Meyer describes the firm's early conservation agricultural projects as "multifunctional beauties," collaborative endeavours with engaged clients that create "experimental grounds for a new kind of sustaining beauty."[7] This skill set has since developed into rigorous specialist expertise with restoration ecologists and conservation biologists who serve as in-studio design collaborators. 3 The Houston Memorial Park expansion is the latest in a wide range of conservation-restoration design projects that go far beneath surface beauty—rooted in pedology, geomorphology, geology, and hydrology to nourish and cultivate the re-establishment of ecosystems as much as sustaining the beloved places they become.

Cultivating Care: A Design Process

In recent years, many landscape architecture studios have necessarily become more attuned to addressing climate change and biodiversity recovery. The development of working methods in ecological design, from the office to the field, is a slow but emerging practice. The profession, bolstered by its new STEM status, has improved opportunities for training, professional development, and certification that draw on the supporting sciences and allied disciplines. Yet much of this remains focused on landscape performance, stuck in the rhetoric of economic benefit. There remains an urgent need for integrated ecological design across research and practice in ways that go beyond landscape performance or ecosystem service metrics. NBW's work, focused on the confluence of ecological and cultural processes, reaches past the service paradigm and embraces a culture of stewardship and care.

The firm has been a steady and early adopter of ecological design that is place-based and stewardship-focused. For more than a decade, the studio has cultivated an in-house interdisciplinary approach to design collaboration. Through staff training, community outreach, and creative collaborations, including community-client partnerships, the studio process is one of deep learning, embedded in site history, place-awareness, and site ecology. Restoration ecologists, conservation biologists, historians, cultural anthropologists, and other specialists are hired as staff members, rather than consultants. Interdisciplinary teams are the norm, not the exception, and this is the basis from which transdisciplinary learning emerges. This radical space of the in-between bridges the arts and sciences, providing a robust basis for an integrated, systems approach to design.

In conversations with staff members, it is apparent the design team shares an explicit working method, not captured in a checklist or a flowchart but rather a way of learning in-and-from place. Projects begin with site immersion, a process of being in and learning a landscape first before design ideas are ever articulated or drawn. A deep curiosity about place is evident, together with a rigorous and structured approach to layers of landscape research. Empirical data and analysis form the basis for an evidence-informed approach. The studio's cultural and ecological teams integrate findings as the basis from which designs are co-created. Listening and learning with community members, clients, and partners, the design team's approach results in projects that are often contemplative, singularly place-focused, and always in the service of care.

Toward an Ecology of Empathy

Good landscape designs are those that cultivate care for the places, people, and cultures they reflect. They cannot be achieved if they are reduced to a checklist of performance or services. Ecological designs are those that cultivate care for nature, for life itself—expressed here as relationship between a place, people, and the other beings that share its lands and waters.

We know that the best hope for a sustainable future lies in nature itself—and in the diversity and abundance of Earth's species, that together provide nature-based solutions for a resilient future—one beyond merely surviving, in which flourishing and thriving are possible. To design such a future means we must embrace the diversity, complexity, and emergence of the living world and its biodiverse species. As we emerge from a global pandemic, human connections to nature and access to ecosystem services are more important than ever. More than half the world's population are now city-dwellers for whom natural ecosystems and their biodiversity still provide essential services (including food, water, pollination, decomposition, carbon storage, climate regulation, etc.) and other resources along with important human health benefits. More than just the essential ecosystem services provided by living landscapes and the biodiversity they harbor, our cities also need the vital benefits of biophilia—the innate human connection to and relationship with nature—one that is increasingly understood to be critical to human physical, mental, social, and cultural wellness.

For much of the world's population, the city may soon be the only landscape our children will ever know. (Re)connecting people to nature in our urbanizing landscapes and growing cities is a critical challenge and a significant opportunity for landscape architecture. Learning to co-exist with nature is part of what it means to be human, yet without this connection, we may lose part of ourselves along with the world we know. To amplify the work ahead, we must engage an ecology of empathy, grounded in awareness and respect for other ways of knowing the world around us. We must learn to see and experience the world through different senses, and learn to design with other species, for a sustainable and flourishing future for all. The design challenge for the polycrisis is to re-engage, re-connect, and re-affirm our connections to nature and to biodiversity. A flourishing future demands an ecology of empathy, from which we design landscapes that re-engage with nature for compassionate co-existence—with biodiversity, with wild and feral species, and ultimately, with one another.

Coda

Landscape, like friendship, is about building a relationship. My relationship with landscape is in part cultivated by my friendship with Thomas Woltz, with whom I have walked through many landscapes—from meadows to mountains, piazzas to prairies. With full disclosure, I am not a distant critic. I have written here from an allied and adjacent discipline, through shared experience, as an honest friend and an accountable colleague. Like the landscapes we have walked together, our friendship has evolved and grown—nourished by curiosity, rooted in shared ideas, competing for the sunlight of time, but well-watered by attention and care (though sometimes dormant through periods of drought). Fifteen years on, through the steady evolution of the NBW design practice, we are seasoned colleagues. In my practice and research, I oscillate daily between conviction and despair. While buoyed and energized by the abundance of interest and emerging talent in ecological design, I grow more dismayed by the urgency of the crises we face. Yet it's now clear that we must make our own hope. In walking the landscapes of NBW, from intimate to expansive, my heart opens. Here lie the seeds of hope, literally and figuratively.

1.
Sense of Place: Meadow/Field – Touch, 2018. Submission by Laurie Greer, Washington DC https://thehaikufoundation.org/a-sense-of-place-meadow-field-touch/

2.
In 2023 U.S. Department of Homeland Security designated landscape architecture a STEM (science, technology, engineering, mathematics) degree program. The American Society of Landscape Architects (ASLA) had previously advocated for this designation. (ASLA, 2023). https://www.asla.org/NewsReleaseDetails.aspx?id=63963

3.
https://naturesacred.org

4.
Claude Cormier, Canadian landscape architect and friend of the author died on September 15, 2023. His landscapes were renowned for their playful whimsy, conjuring joy and wonder. The Cultural Landscape Foundation published this tribute: https://www.tclf.org/canadian-landscape-architect-claude-cormier-designer-innovative-and-playful-public-spaces-dies-63

5.
Wendell Berry, "The Peace of Wild Things." 1968.

6.
https://www.archpaper.com/2023/02/the-land-bridge-at-memorial-park-in-houston-designed-by-nelson-byrd-woltz-opens-with-the-biggest-picnic-in-texas/

7.
Elizabeth Meyer, "Multifunctional Beauties," in W. T. Byrd, T. Woltz, and S. Orr (eds.), *Garden, Park, Community, Farm*. New York: Princeton Architectural Press, 2013, p. 173.

LIFE TO FEED

PROJECT — BROOKLYN NAVAL CEMETERY LANDSCAPE
LOCATION — BROOKLYN, NEW YORK

Before NBW took on this unusual site and made it a meadow, it appeared to be the rare undeveloped parcel within walking distance of downtown Brooklyn. A century earlier, the land had been a burial ground for the Brooklyn Naval Hospital next door, which opened in the 1830s. Both the hospital and cemetery lie within the historical bounds of the Brooklyn Navy Yard around Wallabout Bay, where the U. S. Navy built ships from the early 1800s to the mid-twentieth century. After the government closed the Navy Yard in 1966, the City of New York bought the land, about 300 acres, and eventually a development corporation filled it with film studios and hundreds of other businesses—except for the hospital, which remains vacant, and the cemetery, which is not quite vacant.

In 1926 burials, some unmarked, were removed and taken to Cypress Hills National Cemetery in Brooklyn, though the work was incomplete, as was discovered after the site had become a ballfield. At that point, the grounds fell silent, stranded within a thick shroud of trees alongside a one-way chute of Williamsburg Street at the edge of the Brooklyn-Queens Expressway. There were few appropriate reuses for such a sensitive place.

NBW worked within strict parameters to create the first public space of the nonprofit Brooklyn Greenway Initiative, which has developed a 26-mile bikeway that encircles the borough. Funds for acquiring the land came from Nature Sacred, a national nonprofit that promotes accessible open space in cities as a public health imperative. The group advocates for these urban escapes in part by opening them and in part by compiling evidence, with academic backing, among people that such spaces favor well-being—improved vital signs and metabolism, better mood and fitness.

As for the better mood part, one thing everyone notices about this landscape today is a total bodily shift upon going in. The city outside is a storm of noise and traffic fumes and mazes of buildings and streets. The Naval Cemetery entrance clears a way through that overload, up a low stair to a long, wood-framed gate. A few steps farther in is utter calm, and soon, contemplation. A boardwalk opens before a wide angle of meadow surrounded by a solid fortress of trees. Tall grasses and wildflowers are everywhere, swaying under a dozen breezes at once. The meadow plantings inject a reservoir of pollen- and nectar-rich species into an urban setting mostly devoid of them, bolstering the food web vital both to wildlife, particularly insects, birds, and bats, and to human survival. On a July day, in the heat of noon, there is a blanket of yellow woodland sunflowers and eruptions of purple coneflowers and bee balm. The milkweed is plentiful, but it bloomed in June and is headed for seed.

There is also mugwort, a hated weed, and as likely as not, a Greenway staff person is there pulling it out. Or perhaps serving as a guide. The Greenway hosts regular programs for the community. School groups come to soak in the rich ecology, learn about plants, pollinators, and city wildlife; they watch, listen, draw, or walk around to get to know the place. The heaviest duty of this landscape is to connect people with nature. There are sessions of botanical drawing, meditation, bird-watching, fashion shows, and talks about the nature of nature—a modest amphitheater of benches occupies one corner of the boardwalk.

↑ An axis of granite blocks provides another immersive experience of the meadow.

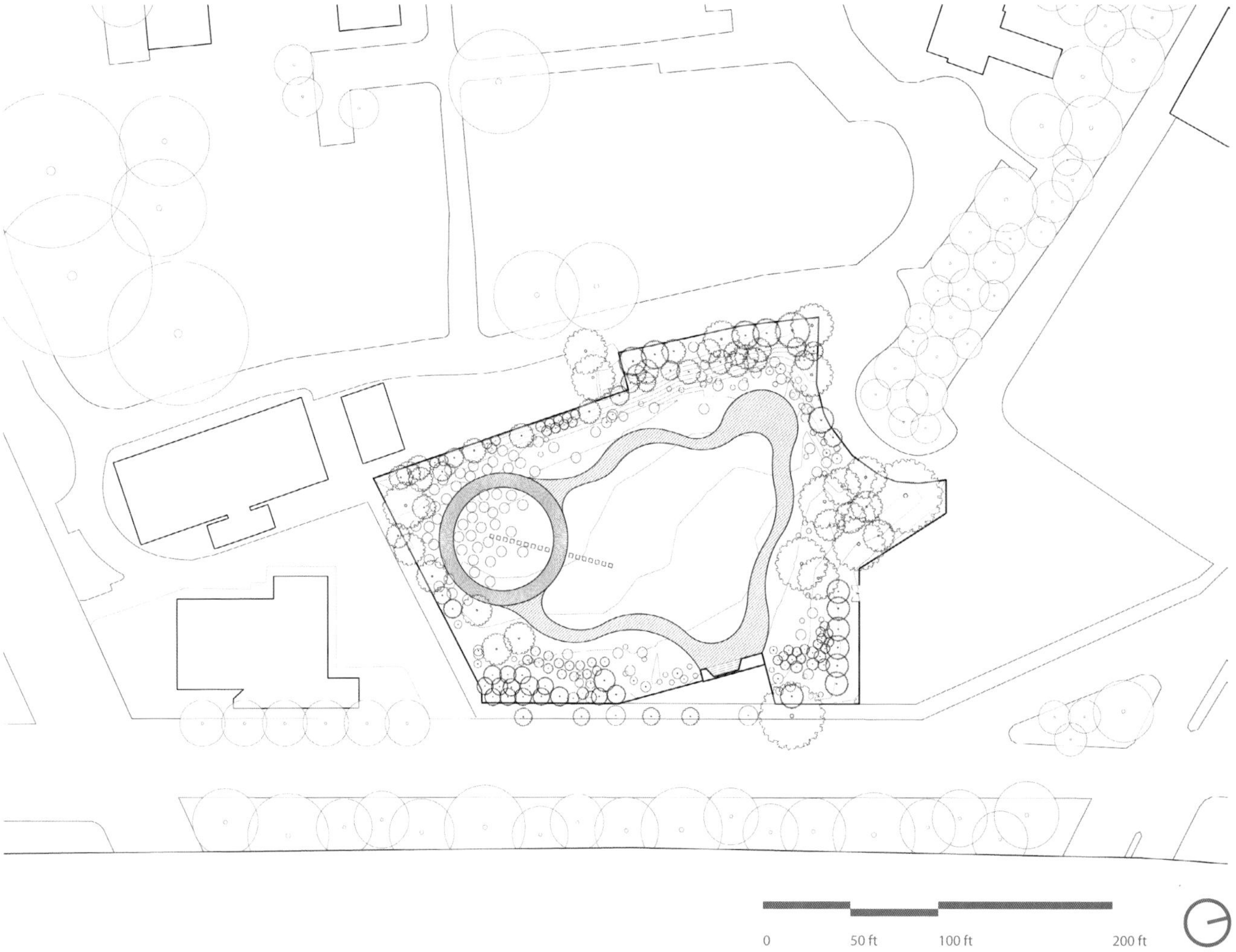

Before the Navy, before the Americans and the Dutch before them, this site was all a wetland in Lenape country. Wallabout Creek spread out and meandered across the site toward Wallabout Bay and the East River. The creek's flows, long since drained out of sight, suggested the shape of the boardwalk of black locust that loops around the site's edges, vanishing beyond curves this way and that, swallowed by the meadow.

The boardwalk floats along its entire path. The firm could disturb the ground no more than four inches down, so the wood structure and entrance threshold rest on a system of surface piers that provide strength and stability without a need for excavation. The constraint on disturbance additionally inspired the major scheme for a prairie of perennial grasses and forbs, the seeds of which could be sown at the surface and pressed in gently.

High granite steppingstones form a broken line out into the meadow—a few long strides to complete immersion. One corner has a miniature grove of black cherry trees, planted small to avoid digging and by now thickened into a screen. Behind them is a bench where Nature Sacred offers a meditation journal in which visitors are encouraged to record their thoughts. Some people sketch. Others confess or grieve. They wish the world well.

There is always life in the meadow. The plantings churn yearly through the greens of ripe grasses and the browns of dormancy in winter. The growing season once again ignites the flat field with the succession of blue false indigo; hot orange butterfly weed; the purples of wild bergamot, hyssop, and coneflower; yellow tickseed and skinny sunflowers, and later, the familiar flush of goldenrod just as blue returns as a fog of asters. The dead receive, as it were, the rites of new and multiplying life. The design hallows the site. The dead, it says, are everyone. They are us.

Wallabout Bay (upper right) and greater Brooklyn depicting dense agricultural land-use, detail of *Plan of the City of New York in North America*, Bernard Ratzer, surveyed 1766–67, published 1776.

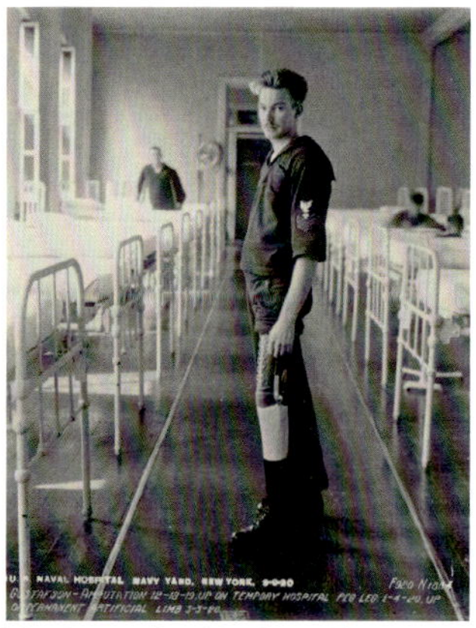

Ward at the Brooklyn Naval Hospital, 1920.

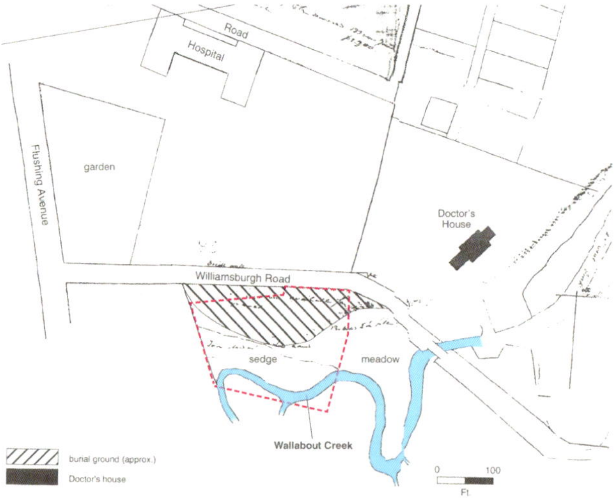

Relative locations of former Wallabout Creek, burial ground, and hospital. Project site in red.

Navy Yard, detail of *The city of Brooklyn*, Charles R. Parsons and Currier & Ives, 1879.

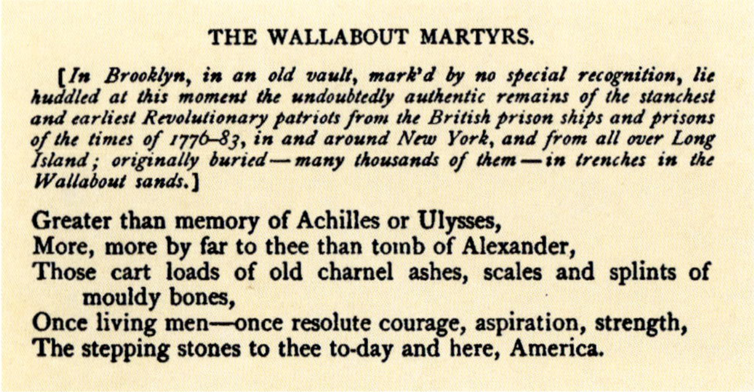

THE WALLABOUT MARTYRS.

[*In Brooklyn, in an old vault, mark'd by no special recognition, lie huddled at this moment the undoubtedly authentic remains of the stanchest and earliest Revolutionary patriots from the British prison ships and prisons of the times of 1776–83, in and around New York, and from all over Long Island; originally buried—many thousands of them—in trenches in the Wallabout sands.*]

Greater than memory of Achilles or Ulysses,
More, more by far to thee than tomb of Alexander,
Those cart loads of old charnel ashes, scales and splints of mouldy bones,
Once living men—once resolute courage, aspiration, strength,
The stepping stones to thee to-day and here, America.

Walt Whitman, "The Wallabout Martyrs," 1886.

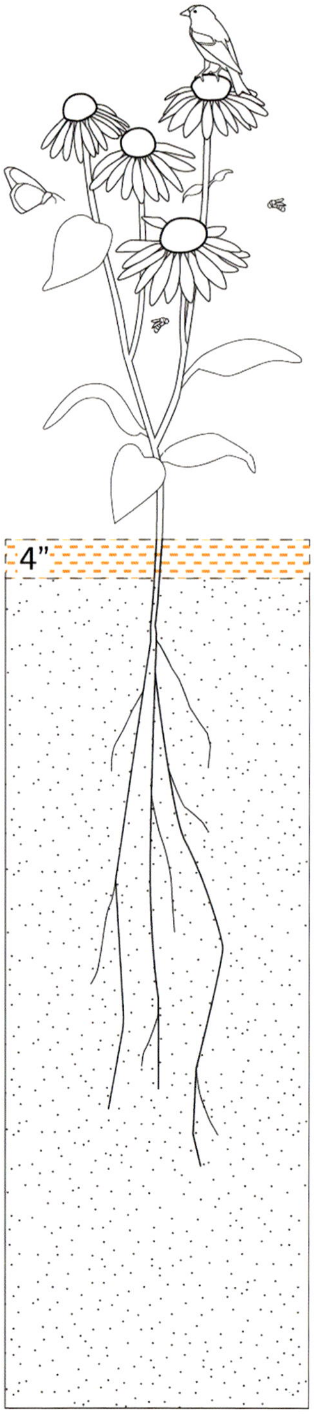

Section of soil depth, meadow planting, and limit of allowed disturbance of 4 inches (dashed at top).

Position of cemetery landscape (far right, near center) as first implemented node of the long-term vision of the Brooklyn Greenway.

↑↑ The floating boardwalk's sinuous form is a reference to Wallabout Creek, which meandered across the site to Wallabout Bay before development of the Brooklyn Navy Yard. ↑ A forested edge encloses the meadow, where native plantings carpet the ground from edge to edge.

↑↑ The meadow opens a refuge of calm in a heavily industrial section of the Brooklyn waterfront. ↑ →→ The boardwalk enlarges around a simple amphitheater of benches, arranged as risers. The black locust planks are set in long lines tangent to the boardwalk's curving edges, with periodic shifts in their orientation.

↑↑ Brooklyn Greenway Initiative staff perform continuous maintenance of the meadow, targeting unwanted aggressive species such as mugwort. ↑ A sheltered bench placed in the black-cherry grove holds a weatherproof journal provided by Nature Sacred, inviting visitors to record their thoughts.

← Bird-watchers have become regular visitors. A yearlong study of birds observed on the site, published in 2021, recorded thirty-eight species with diverse dietary needs, as well as signs suggestive of breeding—exceptional amid severely fragmented urban habitat. ↑ Artist events are regularly staged in the meadow; a fashion show titled "Got Milkweed?" by Collina Strada was inspired by pollinators and regional flora.

SOFT POWER

PROJECT | DUKE UNIVERSITY RECLAMATION POND
LOCATION | DURHAM, NORTH CAROLINA

The shocks of climate extremes are exposing the failure of conventional engineering to contain them, which gives landscape architects openings to reassert their expertise at the center of critical infrastructure design. Machines and heavy engineering in the industrial era created a certain amnesia in infrastructure development about basic solutions that once began in landscapes, such as the gravity-fed ditch irrigation of acequias, the poplar windbreaks across flat farmlands, or the productive heat sinks of south-facing garden walls.

At Duke University, NBW brought de-engineering to a proposal to resolve two major management quandaries for the campus. The result is a pond that captures otherwise fugitive stormwater as a valuable resource and redirects it as a utility. The project adds the obvious benefit of rich, regional ecology for people to experience around a calm mirror of water, which, never mind the utility imperative, could have been its original goal.

The firm took the lead on the project partway through an earlier strategy to carve out a typical stormwater retention pond. Besides capturing precipitation, the pond was intended to supply water to a chiller plant essential to the operations of the campus. The chiller plant serves Duke University Hospital's Level 1 trauma center, which serves a large region and requires copious backup water capacity to avert service interruptions.

A water crisis befell much of North Carolina in 2007. A prolonged drought led the governor to order cuts in water usage across the state. The order restricted household landscape irrigation, which made water scarcity an immediate concern to residents rather than an abstract one. The depletion of groundwater suitable for drinking, irrigation, and raising livestock is a global problem and, in many places, an incipient threat to societies. In the United States, water supplies in arid western regions have long looked tenuous, but in recent years residents of the rainier eastern states are finding their water supplies vulnerable to running low—or dry.

During the drought emergency, Duke University continued to irrigate its sports fields, which invited negative news coverage and harsh criticism in the community. Aware that this usage was far less than that required by the chiller plant, the university sought an engineering solution that would catch rainwater to supply the chiller plant and address the separate issue of storm runoff laden with fertilizer nutrients—nitrogen and phosphorus—used for grounds maintenance. These nutrients in excess promote large algae blooms that rob the water of dissolved oxygen needed by other aquatic life forms. Nutrients were migrating toward a larger watershed from a stream on the West Campus, where the banks had badly eroded. The water held heavy sediment loads from stormwater that had been draining off 120 acres with a high percentage of impervious surfaces.

The engineering proposal for a stormwater retention basin looked perfunctory to Mark Hough, the university's landscape architect. The plan was not aligned with the campus character envisioned by Olmsted Brothers and the university's early architects, Horace Trumbauer and Julian Abele, which has guided campus design since the 1920s. Hough recalled that the Olmsted Brothers plan had specified making at least one pond on campus, which this project could fulfill. He contacted NBW to seek

↑ Duke Pond reduces the campus chiller plant's reliance on potable supplies, allows for the removal of sediments and nutrients in downstream flows, and creates an ecologically productive wetland edge with opportunities for teaching and research.

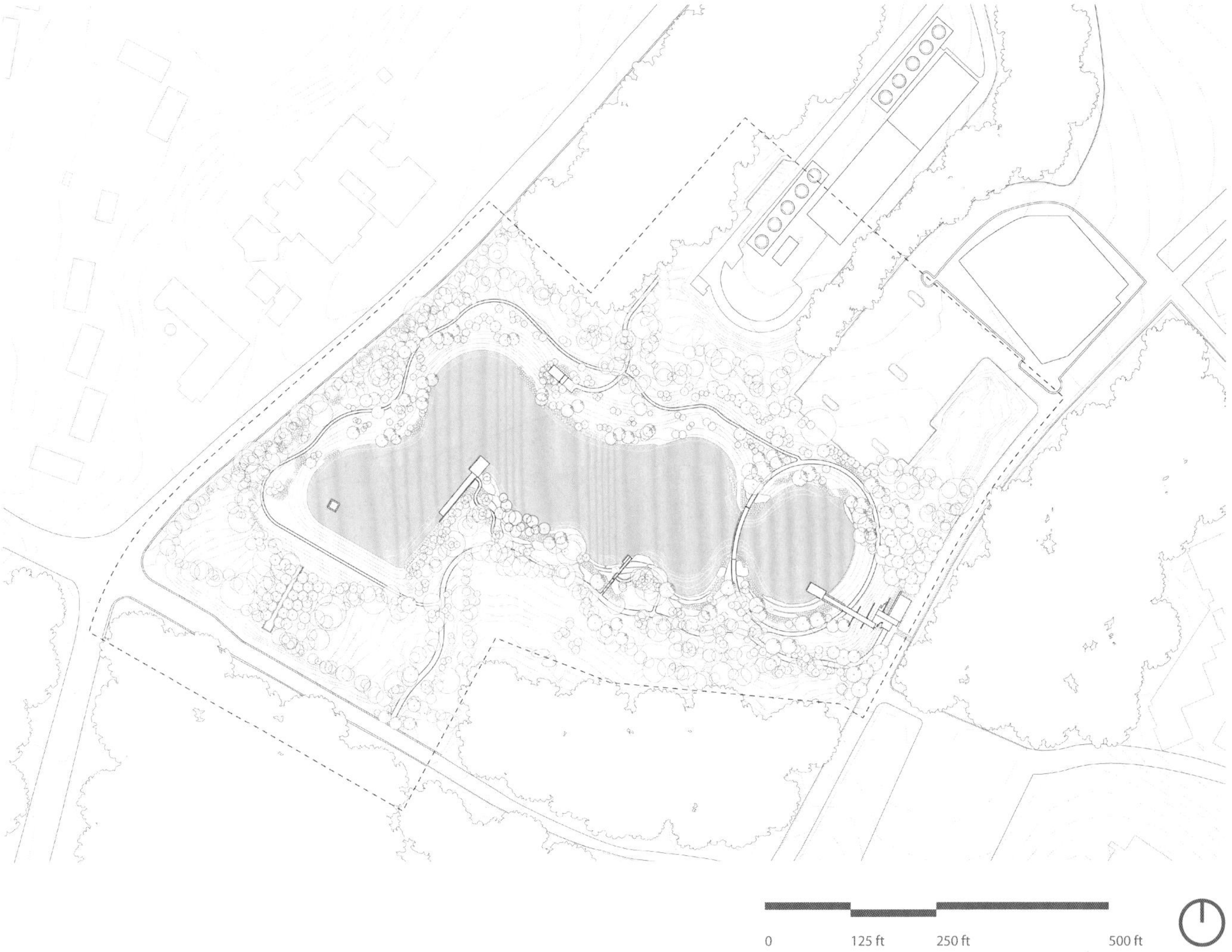

an alternative design. In conversations between Hough and Warren Byrd, the concept turned the retention basin into a living pond much like the Dell, a stormwater mitigation project the firm had completed at the University of Virginia in 2003.

The pond's naturalistic appearance today comes from a push-pull among several factors: making it big enough to supply the chiller plant, ensuring a four-foot maximum flux in the water's elevation, grading the edges precisely to create woodland benches, keeping all trees possible, and planting emergent species that can take inundation and drought. Above all, it would be public. It had to be accessible and a place where people would want to spend time.

On much of the Duke campus, forests and intimate courtyards make for pleasingly cloistered surroundings. The twelve-acre zone around the pond is strikingly different, an open bowl of light and prospect at a sweeping scale that sets a wide angle westward from the overlook above it on Circuit Drive. The view is layered with a front, middle, and beyond: a round forebay enclosed by a footbridge; a widening reach of water toward a pier and pavilion over it, the rest hidden around curves; and a horizon seen through the loblolly pines that spike up on tall, nude trunks.

The design reflects an empathy for the ways people enjoy moving through landscapes, drawn forward by curiosity. A stair at an overlook on stone piers leads down to a path that encircles the pond. The path swerves, gains height above the water, and goes back down closer to the shimmering reflections of green life stacking up across the water. The outfall channel from the dam is clad with Duke stone that defines the overall campus.

The pond has become a living laboratory for students, faculty, and outside researchers, who have found that the ecological health of the zone around it has improved in terms of numerous metrics. Between 2015 and 2019, the Duke University Wetland Center found that the pond had cut nitrogen flowing downstream by an average of 66 percent, phosphorus by 72 percent, and sediments by 94 percent. An in-depth audit conducted in 2020 by an academic team for the Landscape Architecture Foundation made proxy calculations of tree diversity that determined that the pond zone's vegetative integrity rose from below the definitive level for a "natural area" to well above that level. And a qualified count of bird species spotted around the pond and recorded on community-science web applications appeared to show steady rises in new species spotted. The pond has become a draw for people looking to relax or observe nature, and it is also measurably improving the environmental health of the Duke campus and the watershed downstream.

Olmsted Brothers design proposal for Duke campus, showing two ponds.

Interior of the chiller plant.

U.S. Southeast reaches most severe category of drought

Share full article

By Brenda Goodman
Oct. 16, 2007

ATLANTA — For the first time in more than 100 years, much of the southeastern United States has reached the most severe category of drought, climatologists said, creating an emergency so serious that some cities are just months away from running out of water.

2007 record drought in the southeastern U.S. reported by the *New York Times*.

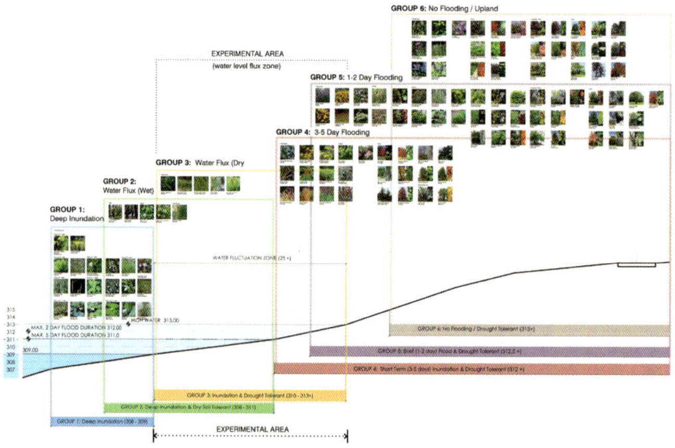

Wetland planting strategy at pond edge for expected water-level fluctuations.

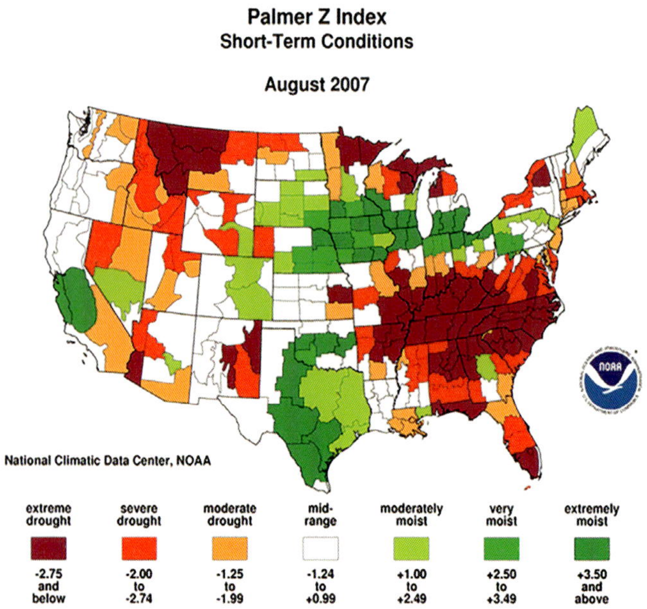

Map of relative U.S. drought severity shows extreme drought in North Carolina, which prompted new water-use strategies across the campus.

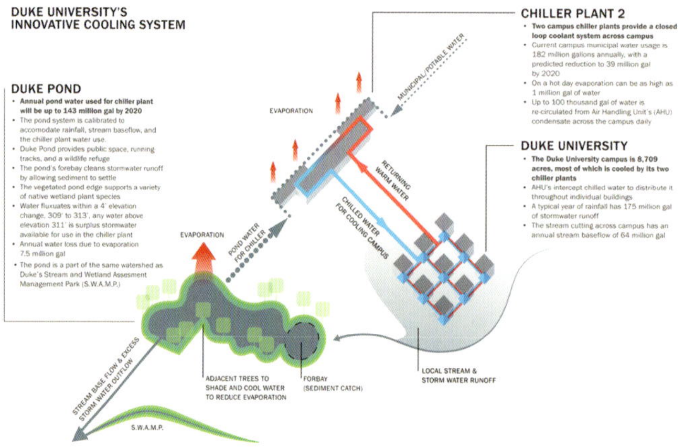

Diagram showing the water/vapor cycle through the pond and chilled water plant.

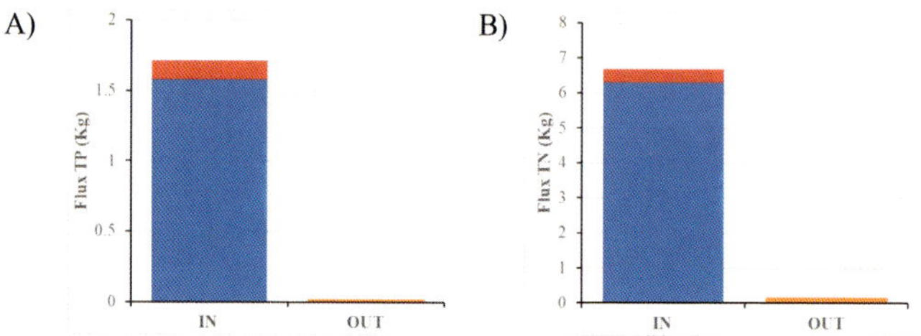

Measured reductions of total nitrogen and phosphorus in the pond during four storms, 2022.

↑ →→ The pond as realized replaced a scheme for a strictly utilitarian stormwater catchment surrounded by fencing. It reflects the Olmsted Brothers vision for ponds in an early landscape plan and has become a focus of research and recreation on the campus.

↑↑ A path along the edge of the pond weaves in and out of the restored Piedmont woodland. ↑ Birds and other wildlife have populated the pond and its surroundings, enriching the campus's biodiversity. ↗ A footbridge crosses the mouth of the forebay. The forebay plays a vital role in capturing sediment runoff that had chronically polluted the stream.

← Mature loblolly and shortleaf pines grow around the pond edge. ↑ Emergent wetland plant communities thrive at the edges, which expand and shrink as water levels fluctuate up to four feet, depending on precipitation and the volume of water drawn by the chiller plant.

CULTURE TO LEAN ON NATURE

PROJECT
LOCATION

THE RAMBLE
FAYETTEVILLE, ARKANSAS

NBW's design of the Ramble creates a ribbon of landscapes and streetscapes that connect several cultural institutions and a series of renewed open spaces in central Fayetteville, the second-largest city in Arkansas, with 100,000 residents. To the north, the Upper Ramble will hold a new civic plaza on a former surface parking lot along Dickson Street, a fine-grained commercial row between downtown and the University of Arkansas campus. The plaza, now under construction, is next to the Walton Arts Center, TheatreSquared, and several other visual and performing arts venues in the cultural center of Fayetteville. To the south, in a wooded stream valley, is the Lower Ramble, the first section to open, which connects to a newly expanded public library, to Tanglewood Branch, and, farther south, to a university satellite hub for arts and design education.

Along the Lower Ramble, extensive restoration work is improving the ecology of Tanglewood Branch and of the surrounding forest known as Fay Jones Woods. A new aerial canopy walk and series of walking and cycling paths let visitors observe up close the layers of nature in the middle of town. This wildish remnant of forest had long been overlooked, whereas the concept of the Ramble is to fold culture and nature together in mutually resonant ways.

The Ramble gained design funding in 2017 from the Walton Family Foundation as part of its Northwest Arkansas Design Excellence Program. Two years later, city voters approved a bond issue to fund its construction. Residents of Fayetteville became vested in the project's outcome early on during workshop-style meetings held to elicit their ideas for this new public realm.

The headwaters of the Tanglewood Branch had been, like many urban streams, in a state of serious decline. About 340 acres of downtown are shedding stormwater into the stream; about 70 percent of that volume resulted from impervious urban surfaces such as streets, parking lots, and roofs. Torrents of water were scouring the earthen valley, eroding soils that clouded the waters, lading it with street pollution, and imperiling trees along the banks.

The polluted stormwater flows were a direct harm to aquatic life in the stream and, in turn, to the health of Fay Jones Woods, a forest named for E. Fay Jones, the renowned architect and University of Arkansas educator who owned the tract before his death in 2004. The original tree population of oaks, hickory, and sycamore in its overstory was flickering out under proliferating nuisance species, the removal of which has begun as a long-term effort, along with replenishing the understory and canopy of resident species.

The work of relieving these pressures on the stream began outside its channel on the surrounding streets. NBW, in collaboration with its Fayetteville counterparts at the landscape architecture firm Flintlock, has redesigned certain streets around the stream for passive stormwater management and greater pedestrian safety. The steep West Avenue, which forms the long eastern edge of the Ramble, has been rebuilt with porous paving and curbside rain gardens as catchments for stormwater before it can run down and overwhelm the stream.

NBW partners at the Watershed Conservation Resource Center completed intensive repairs to the stream and its margins. Revegetation and structural

reinforcements of its banks and upland seeps are re-creating lost habitat around springs and small wetlands and helping to keep the stream's waters cool. The stream channel now has constructed pools, riffles, and rebuilt floodplains to slow the destructive speed of high storm flows and to help flush oxygen into the long-starved aquatic ecosystem.

Currently, the northern section of Tanglewood Branch is in the process of daylighting. Tanglewood Branch originates where multiple springs converge nearby, but it has been piped through a box culvert beneath the pavement for many years. When complete, it will flow within a constructed channel across the Upper Ramble plaza, then convey a short distance southward to flow through the natural stream valley of the Lower Ramble as Tanglewood Branch, onward to the White River and to Beaver Lake, which provides the area's drinking water.

If people in the city had grown to disregard the living environments of Tanglewood Branch, it was mainly because they had few ways to connect intimately with it. A long middle terrace of the stream valley holds a segment of the Razorback Greenway, a 40-mile trail that connects to a network of trails that trace through more than 250 miles of the region. Lateral entry to the valley was limited, and improved public access is the role of the aerial canopy walk, which opens a revealing and dramatic exploration into Fay Jones Woods from West Street directly across from the expanded library.

The canopy walk, which is universally navigable and wide enough for groups to gather at midpoints, spirals among the trees from a high edge of West Avenue, with places for stopping off and taking in the surroundings from wood benches. Its wood railings recall the crafted details of the architecture of Fay Jones.

The path transports people down into the valley gradually, crosses the Razorback path, and continues to broad stone steps that provide access to the stream, which spills with soft musicality on a still day through large stepping stones placed across its width.The adjacent Tanglewood Park feels cloistered and intimate, and connects the canopy walk to the wooded neighborhood across the stream, a unified living theater of nature, now back on a regenerative footing.

↖ An elevated woodland walkway meanders from a street and descends gradually through a woodland canopy before bringing people down to the stream banks. ↑ The linear site extends five blocks and includes 10.15 acres, connecting a civic arts plaza downtown to the restored forested stream valley of Tanglewood Branch and the headwaters of Beaver Lake, the Razorback Greenway regional trail, and a university design center.

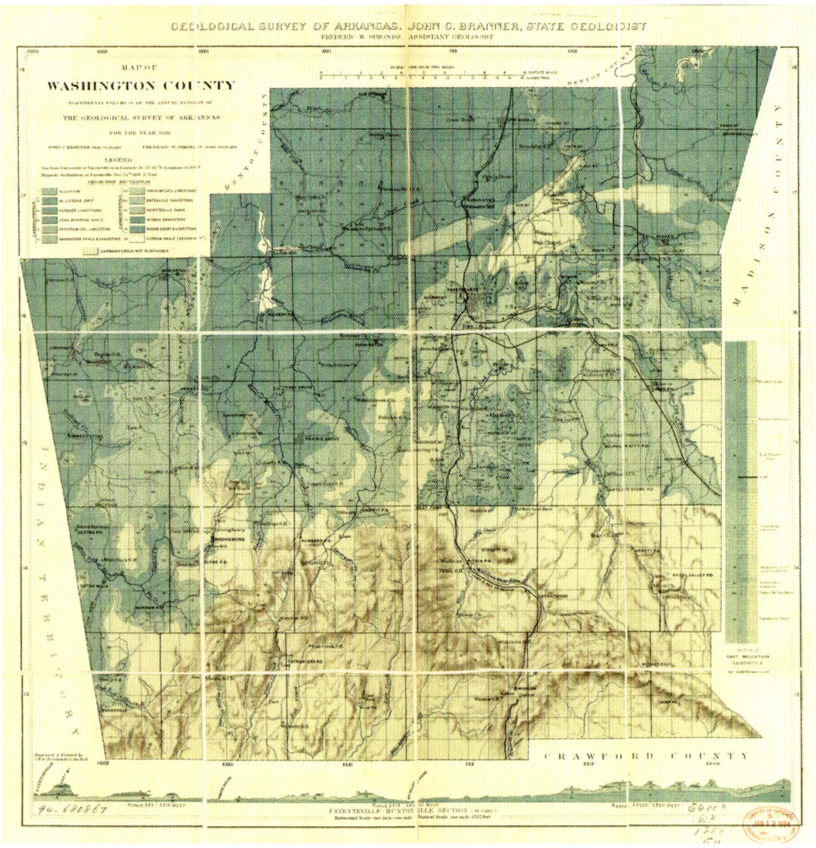

Position of Fayetteville (top row, third square) among diverse formations of sedimentary rock.

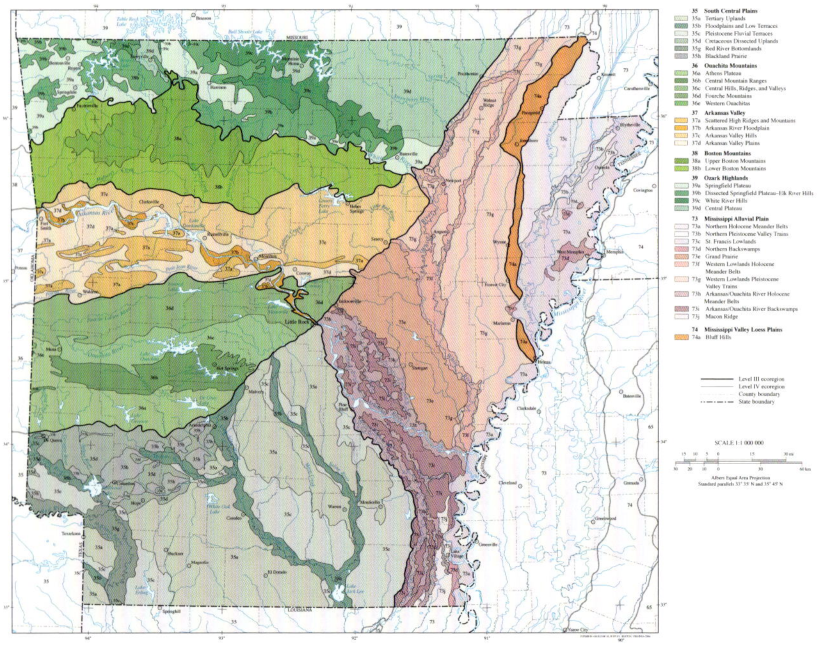

Geological map of Arkansas; Fayetteville is at the boundary of the Lower Boston Mountains and the Springfield Plateau.

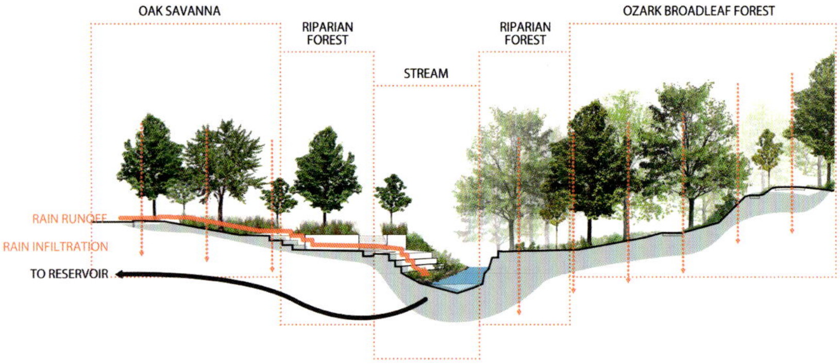

Tanglewood Branch section and map showing ecological subdivisions at Fay Jones Woods.

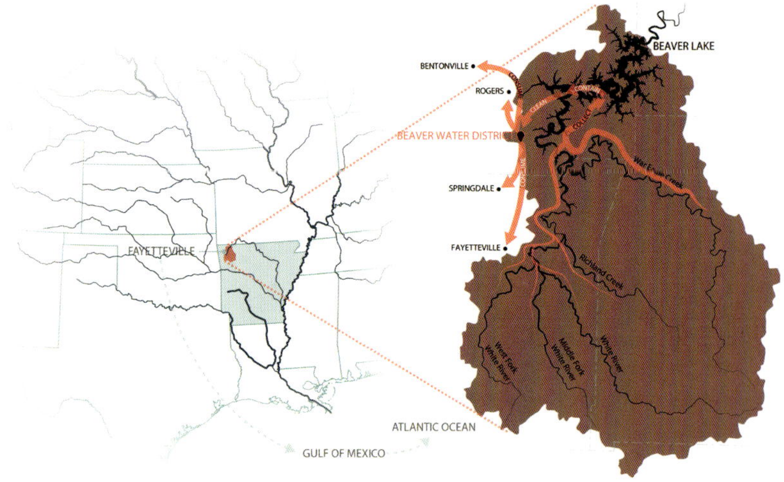

Fayetteville in relation to the Beaver Lake watershed and the Mississippi River basin.

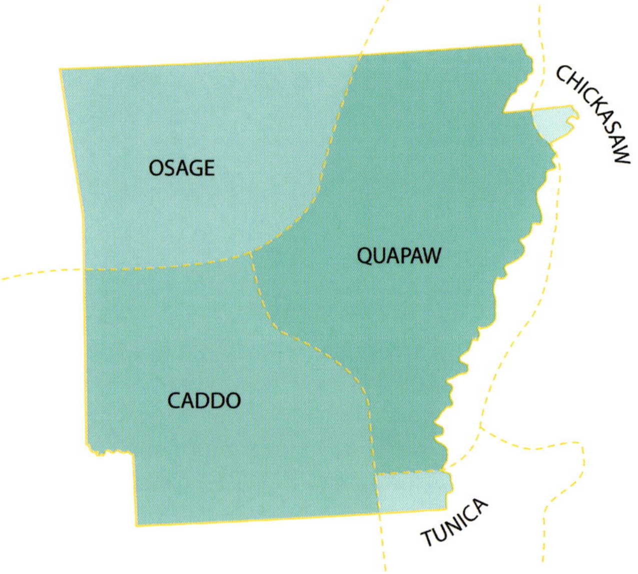

The Osage, Quapaw, Caddo, Chickasaw, and Tunica historically occupied the region aroudn Fayetteville for approximately 14,000 years.

Diagram of The Ramble.

Structural precedent: Thorncrown Chapel, Eureka Springs, Arkansas, E. Fay Jones, 1980.

↑ The canopy walk integrates observation points. →→ The canopy walk path winds gradually down from West Avenue to a seating area along the Razorback Greenway (center) and to the stone terraces of Tanglewood Park by the stream.

The Lower Ramble, set within Fay Jones Woods, opens a variety of experiences for people in groups or seeking solitude within the urban woodland. The path's geometries, generous dimensions, and continuous railings are meant to reassure visitors who might otherwise hesitate to approach a fragment of wilderness in a city.

←Extensive restoration efforts aimed to recuperate the degraded stream, slow the water flow in ephemeral wetlands, and expand biodiversity to help reverse the forest's decline. Stone ledges protect against erosion and sedimentation, and stones placed in the stream channel allow for riffles that oxygenate the water. ↑ Residents and visitors have taken to spontaneous programming of the Lower Ramble with nature walks and a silent discotheque.

III

ESSAY 128 A PURPOSEFUL PURSUIT OF EQUITABLE INTERPRETATION
Brent Leggs

PROJECTS 134 ALETHIA TANNER PARK
Washington, D.C.

146 MACHICOMOCO STATE PARK
Gloucester County, Virginia

164 GEORGIA INSTITUTE OF TECHNOLOGY ECOCOMMONS
Atlanta, Georgia

ESSAY

A PURPOSEFUL PURSUIT OF EQUITABLE INTERPRETATION

Brent Leggs

A step beyond, the dingy streets begin
With all their farce, and silly tragedy—
But here, unmindful of the futile din
You grow your flowers, far wiser certainly.
—Sterling Brown, "To a Certain Lady, in Her Garden"
(for Anne Spencer)

History is a revelation. We can dig deeper. We can uproot, turn over, and extract truths from our landscapes, architecture, and places. Through historic preservation, we restore connections to place and reclaim the work of preserving memory, helping us make sense of who and where we are. Not only do preservationists and design professionals play a unique role in discovering and exploring untold histories, they also have a hand in creating culture and community at historic sites through love, joy, and creativity.

In his book *Black Landscapes Matter*, the landscape designer Walter Hood says, "Black landscapes matter because they are renewable . . . We can uncover, exhume, validate, and celebrate these landscapes through new narratives and stories that choose to return to us as origins." In the spirit of this concept for an expanded American story, the African American Cultural Heritage Action Fund—a program of the National Trust for Historic Preservation—was created in 2017 as a response to the events in Charlottesville, Virginia. Today, the Action Fund is the largest preservation effort ever undertaken on behalf of African American historic sites and the organizations and institutions that support them. It is revolutionizing the U.S. preservation field with its dedication to securing permanent representation for Black history at a time when it is being suppressed. We are igniting a cultural reckoning, shifting our nation's understanding of its history, and building an ethic for heritage conservation. By elevating stories and places of African American resilience, activism, and achievement, we are making lasting contributions to our nation's cultural landscape. To date, the Action Fund has provided grants to more than 297 historic African American places and invested more than $27 million to help preserve landscapes and buildings imbued with Black life, humanity, and cultural heritage.

The aspiration is that preserving these sites and telling their stories will spark the public's curiosity, encouraging people to visit, learn, and take inspiration from these historic locations. Everyone should be able to see themselves and their communities in these stories, and visitors should encounter knowledge they can bring to conversations with their families and neighbors about a richer, more complete telling of America's history. As design and cultural heritage professionals, we are responsible for using preservation and the restoration of our collective narrative as a force for positive change.

In the not-too-distant past, historic preservation favored Eurocentric narratives and buildings over their environmental contexts. However, we have been reminded through our partnership with the

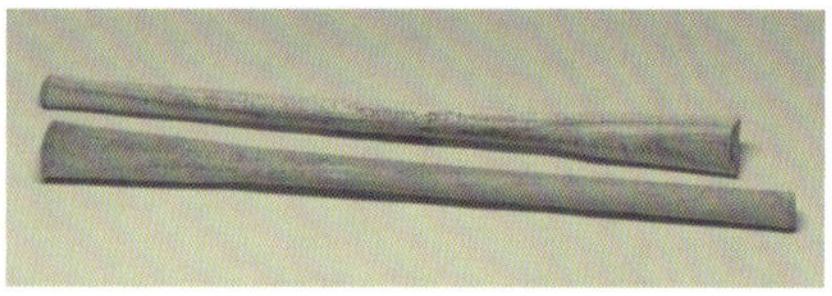

fig. 1 Pickrick Drumstick signed by Lester Maddox, c. 1964.

landscape historian Thaïsa Way, "Don't forget the landscape and the diversity of stories it can harvest." Black people's historic imprints within landscapes have routinely been rendered invisible, particularly in examples of vernacular architecture, with its embodied beauty, representation, and deep ties to the land. We can mitigate erasure by recovering and integrating landscape architecture with historic preservation, thereby interpreting the prolific history of the Black experience.

Landscape architects bring technical expertise, and they also employ preservation techniques to find connections between architecture and landscape design. Through commitments to equity, documentation, and interpretation, engaging the land itself can open up new ways to commemorate lost or hidden stories. There are obscured layers of history waiting to be unearthed, but they require, as Robert Pogue Harrison writes in his book *Gardens*, "releasing the art of seeing and reaccessing the deep time folded within their forms."

When I first encountered Nelson Byrd Woltz's work a decade ago, I was impressed by their holistic understanding of historic preservation and their commitment to equitable interpretation. "We don't know what to do," Thomas Woltz told me, referring to his NBW design colleagues, "until we ask the land its story." The firm encounters this opportunity with regularity, including with projects that start as something else, something more straightforward.

Among the first of these sites that caught my attention was a new campus open space at Georgia Institute of Technology. The school's new EcoCommons is eight acres of wild meadow, part of eighty interconnected acres of related projects that weave native Piedmont species through the campus, a register of ecological memory impenetrably rich with regional plantings.

However, at one edge, a space opens up and creates a sensory shift as you enter a commemorative site. The campus holds the largely forgotten footprint of The Pickrick restaurant, torn down in 2008. The Pickrick was a segregated establishment where, in July 1964, three Black seminary students from the nearby Interdenominational Theological Center—Albert Lee Dunn, Woodrow T. Lewis, and George Willis Jr.—attempted to dine. Its owner (a future Georgia governor) refused to seat them, inciting a mob that chased the three men out with baseball bats and guns. (1) This led swiftly to the first petition to the Supreme Court to uphold the Civil Rights Act of 1964 and affirm a guaranteed right to public accommodations.

NBW came upon the diner's existence when viewing old property records during their site research and parlayed its vanished imprint into a place of interpretation and honor. The design's elements represent barriers formidable and then broken, the determination of the three students, and the vindication of their struggle. It reconnects the stream of its past to the present once more, demonstrating how landscape design can restore the narrative flow of a place. The transformed space now redefines its history as one of inspiration and courage for students, faculty, and visitors in the present.

At another NBW project, in Texas, the winding pathways in

fig. 2 Camp Logan, c. 1917.

fig. 3 Aerial view of the Monticello mountain top; The Contemplative Site marking the 607 people held in bondage by Thomas Jefferson, with flowers left by visitors; Entrance to the Burial Ground for Enslaved People.

fig. 4 Eckington Yards, 1923.

Houston's Memorial Park pull visitors forward, inviting them to wonder about what is ahead. NBW's master plan began with a major focus on turning around an ecological emergency in the park's dying forests, but through the firm's research-based approach, they also explored its historical and cultural context in depth. The project is notable for its constructed, meadowed land bridges over a roadway that allows for the safe passage of wildlife and people.

In addition to being Houston's largest park, it is also the former site of Camp Logan, [2] where, in 1917, racial tensions boiled over between Black soldiers guarding its construction and white Houston police and residents. The incident that followed left seventeen dead and led to the largest court-martial and murder trial in U.S. history, with thirteen soldiers hanged. During the park's present-day redesign, the landscape architects hoped that the new structures, site features, and fixtures would allow visitors to see Camp Logan as a sober reclamation of the turmoil of its past. Thanks to encouragement and continued engagement by NBW, there is an opportunity to integrate Black history and interpretation across this sublime landscape while creating a new section of the park.

Using similar methods, the firm made a discovery at Monticello, Thomas Jefferson's 5,000-acre plantation in Charlottesville. [3] Paths appeared on a survey of Jefferson-designed roads across the site that had become obscured and covered over time. One was used for carriages, but enslaved laborers used another to tote water up and down the hill. "These were the people who were literally supplying life to the house," Woltz said. The research was used to re-create those paths, centering their story and those of their descendants, giving form to the Monticello Burial Ground for Enslaved People, a quiet, sacred space for contemplation among the trees.

Place, and its specificity, means everything to this work. Alethia Tanner Park in Northeast Washington, D.C., is on the former site of the Eckington Yards, [4] a freight hub of the Baltimore and Ohio Railroad—a connection point for rail workers and passengers alike. Alethia Tanner Park differs significantly from NBW's other projects, as it acquired its name after its completion. Woltz believes that the initial opportunity to integrate Tanner's story into the park itself was lost. "How do you tell a story that doesn't belong here?" he said. Tanner is more aligned with Lafayette Square, the park directly in front of the White House, where she sold produce and eventually bought her freedom and that of her family. Inspired by this incredible journey from enslavement to freedom and accomplishment, Woltz described how he would have envisioned the flow of the timeline of her life, winding through the park like a river, mirroring how she moved through the world—where Tanner and her family came from and where they went. "There's a way to demystify design. To make it feel approachable and convey meaningful work," Woltz said. "I want history to be energetic. Like a spring—gushing forth."

Because the railroad polluted the site's soil, the park's plantings now include meadow grasses, such as blue-eyed grass, and trees whose roots will break down the toxins within the soil over time, with the

fig. 5 Nina Simone's childhood home in Tryon, North Carolina.

fig. 6 Stills from *Chasing Trane: The John Coltrane Documentary*, 2016.

fig. 7 Album cover for John Coltrane's *A Love Supreme*, 1965.

hope of long-term environmental remediation effects. The same could be said for the practice of restoring history. We must break down misinformation and erasure by allowing untold truths to take root. It will take time, but if we cultivate them now, eventually these histories will grow and counteract toxic beliefs, allowing us to reshape and restore the narrative.

After observing the progress of NBW's landscape projects over the years, the Action Fund has partnered with the firm to share how history shapes our understanding of place and to develop innovative approaches to preserving and interpreting Black cultural heritage sites. We have always appreciated how NBW and its talented roster of strategists, design thinkers, and cultural documentarians bring added layers of depth to our projects, specifically the restoration of Nina Simone's childhood home in Tryon, North Carolina, and the John and Alice Coltrane home in Dix Hills, New York.

In the southwestern North Carolina town of Tryon is a three-room, white clapboard house where a young Eunice Waymon fell in love with music, setting the stage for the icon she would become—the jazz musician and activist Nina Simone. 5 The vernacular building embodies a conversation about architecture and the landscape, about legacy and the truths of an African American family's life in the rural South of the 1930s and 1940s. The 560-square-foot home was saved from the threat of demolition in 2017 through the intervention of prominent African American artists Adam Pendleton, Julie Mehretu, Ellen Gallagher, and Rashid Johnson. Pendleton learned in a phone call that Nina Simone's childhood home was for sale and did not want its story to become lost. The Action Fund manages the project via a partnership with the four artists, local community members, the World Monuments Fund, the Nina Simone Project, the North Carolina African American Heritage Commission, Pace Gallery, and others to ensure that Simone's home and legacy is protected in perpetuity as a monument of pride, progress, and creativity.

NBW was invited to design the home's grounds using an approach that reconnects it to vernacular traditional and cultural landscapes through elements such as a reimagined swept yard. The design calls for historical aspects of the house to be interpreted through simple materials and plantings, such as a frame of bricks to mark the footprint of a former shed. These reflect the humble beginnings and roots of what would influence Simone's iconic future.

Similarly, the home of John Coltrane and Alice Coltrane 6 in Dix Hills, New York, is a familiar enough domestic setting that contained extraordinary creativity and invention. Through the Action Fund, the NTHP has been supporting the stewardship of the unassuming brick ranch house, built in 1952, that conceals the incredible musicianship that took place there from 1964 to 1973. It was here that John Coltrane wrote his 1964 masterpiece *A Love Supreme* 7 in an upstairs bedroom. In her basement studio, Alice Coltrane recorded five albums, including her 1968 album *A Monastic Trio* and, with John, *Cosmic Music*.

The house and grounds were saved from demolition in 2002 and are now owned by the Friends of the John and Alice Coltrane Home. In 2020, the NTHP commissioned NBW to design the landscape plan for the 3.4 acres surrounding the house. For Alice and John Coltrane, creating and sharing their music was an expression of a continuing evolution—not just as musicians, but of their spiritual selves. A future project for the site involves the creation of a park and walking trail at the rear of the house inspired by John Coltrane's manuscripts for *A Love Supreme* and its innovative jazz progressive chords and reharmonization. The park will also have a meditation garden to channel the spiritual nature of the Coltranes' lives and work.

Across all these projects, professionals at the Action Fund, the NTHP, and NBW advance equitable site interpretations that uplift the Black American experience and story. The Action Fund serves to make the hidden histories of Black America visible through stories of achievement, activism, and resilience. The focus is not solely on the material fabric of a particular site, but also on the intangible aspects that contribute to a full history of place. At Action Fund–supported sites such as Faith Ringgold's Home & Studio in Englewood, New Jersey, the Anne Spencer House & Garden in Lynchburg, Virginia, and the Sojourner Truth Legacy Plaza in Akron, Ohio, landscapes are being transformed into places of education and new concepts for community engagement. It is more than restoration of the physical environment; it is the restoration of memory.

Good design can resurrect the past and envision new connections between history and landscapes for our country's shared future. Current and emerging professionals in our field are developing a deeper appreciation for the restorative power of nature, our connection to it, and how we can communicate and represent these collective values in public spaces. Part of our ongoing work is helping communities and institutions access and implement design solutions to preserve cultural heritage and the environment and to tell the full American story. In doing so, we hope to continue engaging partners like NBW that are committed to expanding a pipeline for diverse landscape design professionals to preserve these places and inspire a commitment to racial, social, and environmental justice.

TIMES AND LIVES

PROJECT
LOCATION

ALETHIA TANNER PARK
WASHINGTON, D.C.

NBW won a competition in 2016 held by the NoMa Business Improvement District, through its parks foundation, NoMa Parks, to design a 2.5-acre park in the Eckington section of Northeast Washington, an area that has redeveloped rapidly but lacks park and play space. NoMa Parks wanted spaces for passive recreation and community events. Public engagement meetings elicited the hopes of residents for a safe, well-lit park. NBW's design responded also to wishes for play spaces for kids and pets, a lawn ample for performances or movies, plentiful seating, and shade trees. The site, a former rail yard, had legacy pollution in the soils, which NBW removed or contained and otherwise addressed with heavy plantings of species that lock up soil toxins. In another public participation process, the community voted to name the park after Alethia Tanner, an entrepreneur and civic leader among early Washington's communities of enslaved and freed persons. In 2022 two of Tanner's descendants attended the park's dedication in her name. Visits over the following year revealed the range of attachments residents have formed with the park.

Tuesday, March 1
10:30 a.m.

The workday morning was bright and silent across Alethia Tanner Park. A few people were around, power-walking or being towed by leash to the dog run. Amtrak and Metrorail trains crashed through the quiet. The tracks run just over a high boundary wall several hundred feet long, which is covered end to end by a series of wildly colorful murals. On the main entrance promenade, dozens of café chairs awaited the off-hours of a warmer day, arrayed beneath a large metal canopy that stretches off a small wood-clad pavilion the color of charcoal. The forms and textures of the railroad show up in those of the park. There are benches and chaises all around made of wood faded to gray. At the promenade's edge is a high black steel frame, like a gantry—a proscenium for a screen on movie nights—and a deep central lawn that spreads out with room for about a thousand people.

Saturday, June 17
1 p.m.

Solstice warmth—low 80s, a little sticky; the shadows right under everything. In good weather the park is Eckington's patio. People relaxed alone or in clumps, mainly in the shade. A DJ was setting up under the gantry, which holds a huge banner, black on yellow: ALETHIA TANNER PARK. Panels nearby tell the story: Alethia "Lethe" Tanner was born in 1781 into enslavement nearby in Maryland. She became a produce merchant next to the White House and then bought freedom for herself and twenty-three family members. She laid foundations for Black education and worship in Washington that endure today.

Eckington is locked amid three commuter boulevards up against the rail corridor's last mile south to Union Station. The park sits well back from that chaos of cars, buses, and trucks, and its prodigious plantings offer insulation. There are trees arranged as allées, high hedges, groves, and specimens. And on the edges nearest the existing tracks, huge grasses, rushes, and wildflowers fill bioswales that capture stormwater before it flows to the sewer.

For the moment, stormwater was not a problem—there had been no storms. The central lawn had turned brown. People were drawn to it anyway.

Sunday, August 6,
4:30 p.m.

Rains had come at last. The central lawn was peak green and soft. Facing the lawn, on a bench, was Delores H., ready to engage. Every day, Delores said, she leaves her apartment, crosses the street, and walks two or three times around Alethia Tanner Park. Despite the August heat index of 95 degrees, today was no different. She knew several people who walked by and caught up briefly with them. Delores turned and said, as an aside, "I'll come over and have a conversation out of the blue for no reason."

Thursday, August 17,
7:30 p.m.

A weekly farmers market was set up along the allée, with crates of peaches and flats of beans, berries, and peppers. Nearby, a duo was playing jazz. A volleyball game on the lawn erupted with cheers and groans. There were bicycles everywhere on the ground, against poles, rolling down the regional bike trail along the murals. Just after 8 p.m., several dozen people coalesced from the scattered population at a front corner of the park to begin an informal memorial observance for someone in the neighborhood. "We are here because we Eckingtonians come together," a speaker began. "We celebrate together. And we grieve together." It was near dark. People lit candles and leaned on each other.

Wednesday, November 22,
3:45 p.m.

An hour before sunset, decks of bruise-colored clouds pushed through bright blue skies. The shadows of everything stretched long. In the partial gloom, the sun supercharged the park's autumn colors. Elms studded along the allée glowed like torches, and the colors of the murals appeared electrified by an oblique shaft of sunlight along the bike trail. The meadow colors, green and boiling over the bioswales in August, had turned tan and brown toward winter, slashed by the red blades of little bluestem. A boardwalk runs the length over one swale. A toddler on training wheels clattered over the planks; her mother kept pace on foot and kept an amused watch.

As twilight came on, three men winding down a day of park maintenance began to talk about the park. They seemed uncertain who Alethia Tanner is or was, and so were directed to the short bio of Tanner on a panel nearby. They walked over, and one read it aloud. He reached the final lines: "Alethia Tanner died in 1864 after a long life in which she witnessed the creation of the nation's capital," he recited, "gathered the means to help herself break free from slavery, and became a founding member of Black institutions that still exist today." None of them said a word at the end. None moved an inch for several seconds.

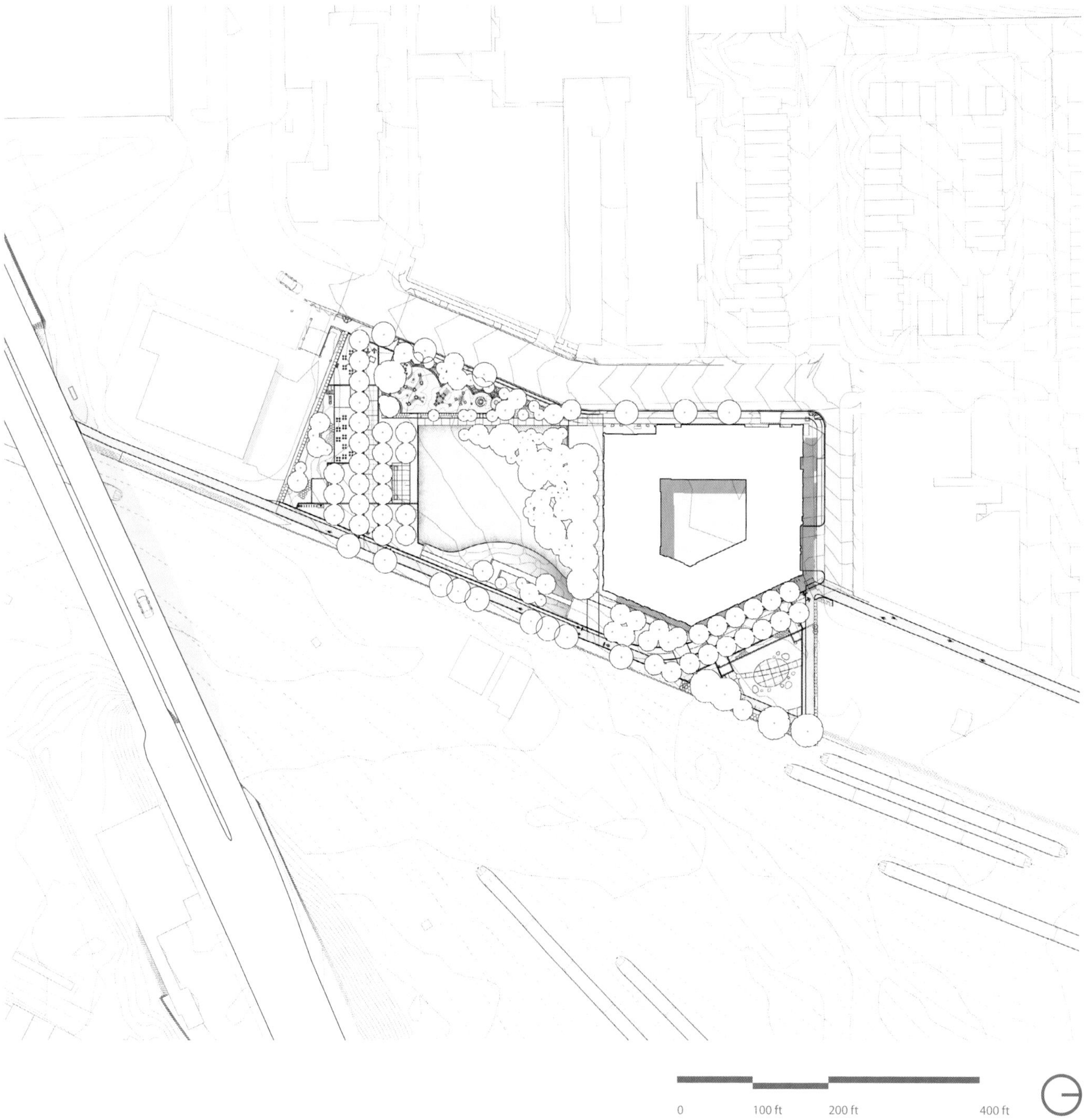

← The central lawn can hold about 1,000 people on movie nights in spring and fall. ↑ The park occupies a former freight rail yard in a neighborhood undergoing intensive redevelopment. To the east, beyond a a segment of Metropolitan Branch Trail that forms its boundary, working rail lines carry local, regional, and cross-country trains.

Washington, D.C., property valuation map, 1892; B&O rail yard is in the triangle center right.

Research graphic showing Alethia Tanner Park located in an area without many key public amenities, especially public parks.

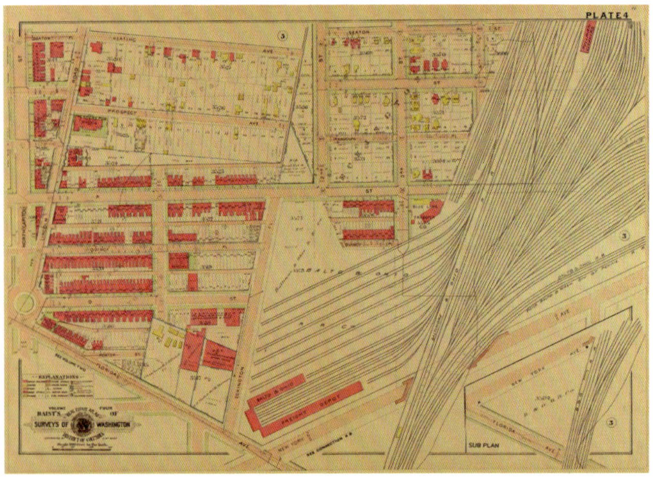

Survey plat of Washington, D.C., showing the recently constructed rail yard, 1907.

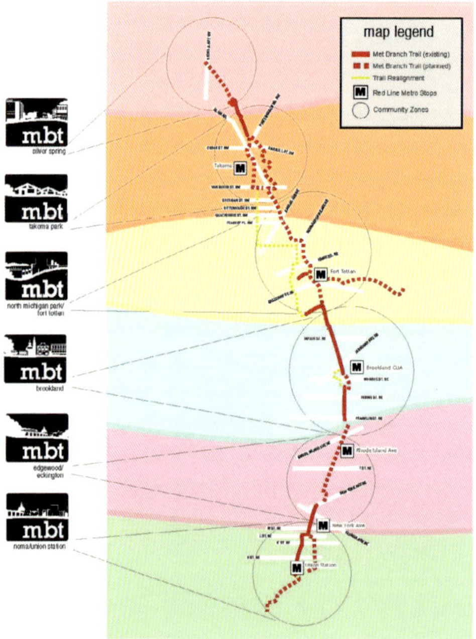

Schematic diagram of Metropolitan Branch Trail; the park site is in the pink zone.

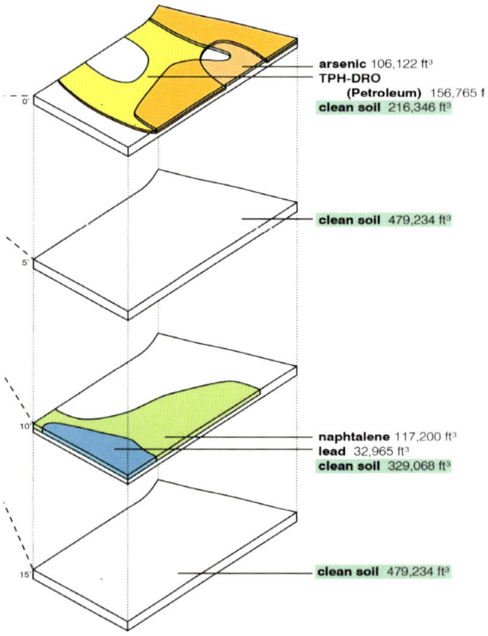

An exploded analysis diagram illustrates relative locations of pollutants from rail operations.

U.S. Capitol dome seen from the rail yards in Southeast section, c. 1917.

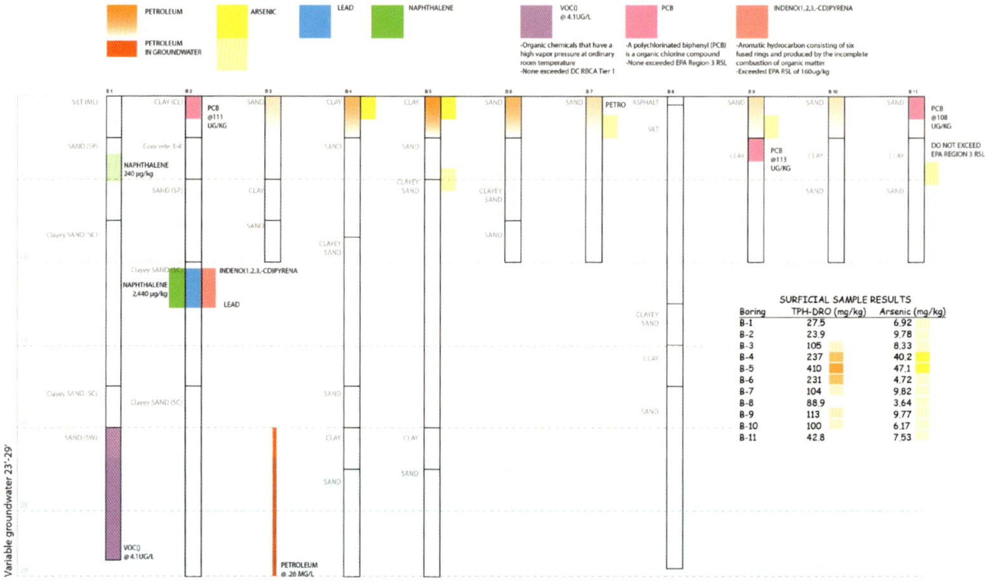

Soil borings showed the depth and extent of hazardous wastes to twenty-nine feet below surface. The contaminated soil was kept on-site and is undergoing photoremediation clean-up by plants. Interpretive signage explains the process to visitors.

↑↑ A boardwalk floats across an area of wet meadow designed for the phytoremediation of legacy pollutants left by the site's use as a rail yard. ↑ Lively programming of movies, concerts, festivals, and a weekly farmers market bring large numbers of visitors to the park throughout the warm seasons. →→ A segment of the Metropolitan Branch Trail, a regional bikeway, borders the park.

↑↑ Serving a diverse community, the spaces of the park were designed to be flexible and accommodating of varied daily uses. ↑ Seating at the north plaza offers multiple gathering options. The space is often programmed with community-led workshops. → The multi-functional structures of the stage and cafe pavilion on the south plaza frame space for crowds.

Welcome to
ALETHIA TANNER PARK
in Eckington
NoMA
SPONSORED
FREE
EVENTS
NOMA NIGHTS: TUESDAY
CONCERT SERIES
SPRING/FALL
CiNoMatic
NOMA FARMERS MARKET
NOMA in BLOOM
BEER TRAIL
PUMPKIN PALOOZA
NoMaBID.org/Events

← The playground's physical and experiential attractions are inclusive of kids of all abilities. ↑↑ An annual mural festival invites local artists to paint new panels on the 900-foot-long boundary wall. ↑ The mural wall follows the Metropolitan Branch Trail.

THE LONG ARC

PROJECT LOCATION

MACHICOMOCO STATE PARK
GLOUCESTER COUNTY, VIRGINIA

When visiting Machicomoco *(Wingapo! [Welcome!])* in the Tidewater region of Virginia, people gather a palpable sense of history, or historicity, in the landscape. The guided movement through and along the varied features of the park attune visitors to the legacy of the more than thirty tribes whose people inhabited the Coastal Plain for at least 18,000 years. Over many successive generations, these tribes developed a deep knowledge and relationship with the environment—the flora, fauna, earth, and water. Only ten miles up the Pamunkey River (now also called the York River) is Werowocomoco, a sacred site for the Virginia Algonquian tribes over thousands of years. At the time of English contact, the tribes had been united under the paramount Chief Powhatan (Wahunsenacawh), and Werewocomoco had been his seat of power. In 1607 Chief Powhatan (Wahunsenacawh) and Pocahontas encountered the arriving English captain John Smith, and history took a turn.

Revealing the long history through the medium of landscape architecture becomes the critical challenge. Research and records help, as does the land itself. There are the knowledge keepers, the Indigenous people who sustain this presence, and they are critical informants to the scholars, workers, archivists, and archaeologists whose expertise is dedicated to the recording/archiving of this history. They overlap in conversations, meetings, at shared meals and while walking the land together, observing the shell midden and the eroding coastal edge that is giving up its artifacts.

The NBW process arranges for this to happen through research and dialogue, with space to assess and then respond with respect in hearing the stories and understanding of history. The design for Machicomoco was created in collaboration with a team consisting of Virginia Algonquian Tribes, the Virginia Department of Conservation and Recreation, and faculty from the College of William and Mary. The focus of the plan was to reveal the Virginia Algonquian landscape artfully and compellingly to engage a wide audience with this history. The strategy for this communication relied on two devices: one that provided information on the Virginia Algonquian tribes' past, present, and projected future, and another that sought to provide physical access through the landscape and to the water in ways that would rely on this natural environment to help visitors understand the Algonquian worldview. In a sense, it meant decolonializing the landscape.

Meetings with Algonquian tribal leaders and stakeholders ensured that interpretive elements were grounded in truth and honored community histories, identities, and aspirations. The tribe representatives who became partners include: Chief Stephen Adkins (Chickahominy); Chief Walter David "Red Hawk" Brown, (Cheroenhaka Nottoway Nation); Chief Carl Custalow (Mattaponi); Chief Dean Branham (Monacan Indian Nation); Chief Lynette Allston (Nottoway Indian Tribe); Chief Robert Gray and Dr. Ashley Atkins Spivey (Pamunkey); Chief John Lightner (Patawomeck Indians); Chief Anne Richardson (Rappahannock Tribe); Chief W. Frank Adams (Upper Mattaponi Tribe); Chief Samuel Bass (Nansemond Indian Nation); and Chief Gerald Stewart (Eastern Chickahominy Tribe).

The team also relied on research by archaeologists and scholars with expertise on this history; Professor Martin Gallivan and his team at William and Mary were especially crucial to the process. Paramount to the design was a focus on Algonquian agency and the imagining and celebration of a rich and thriving Virginia Algonquian future. The landscape

architects' immersion in the process of archaeological discovery drove the experiential design. In one instance, the discovery of a shell midden on the banks of the Pamunkey River served as a key clue to Virginia Algonquian lifeways and also redirected the creation of a kayak launch to a less sensitive location.

The design utilizes natural materials throughout the park in ways that engage the dynamics of the site, from weather to maintenance. An abstracted block of oyster shells marks the way into the park with a clear line toward the Pavilion. The simple cubic form alludes to the human ability to abstract and idealize form, as well as a visceral reminder of the power of the land and the stories it holds. Past the shell midden block, the procession to the Pavilion follows a historical timeline focused on the people of Tsenacomocah and the geological and hydrological events that led to the initial occupation of the Chesapeake Bay region. The text describing these events was carefully curated with tribal leaders to ensure accuracy and to address challenging ambiguities in building effective and meaningful narratives.

The timeline extends into the Pavilion, a structure inspired by Algonquian longhouse forms. Within the Pavilion, the timeline marks the point of contact with European colonialists, the ongoing existence and persistence of Algonquian communities, and then continues out, projecting toward their unfolding future. The design provides the potential to install new interpretive material to commemorate future milestone events.

Inside the pavilion visitors find a series of interpretive panels that briefly introduces key aspects of Algonquian History and Identity. The panel content is organized into six categories: The Algonquian Landscape, History and Community, Managing a Dynamic Landscape, Placemaking and Mobility, Authority and Ceremony, and Persistence. Each panel illuminates histories and lifeways of regional Indigenous populations that have historically been ignored or hidden, honoring them in the context of the land.

Simple but distinct symbology was developed, which in conjunction with key Algonquian terms, developed into graphics that are applied sensitively throughout the site, on structures, and other elements in the landscape to connect visitors to specific. These cues fold seamlessly into the site design and remind visitors of Algonquian history throughout their time on the site without over-complicating aesthetics or distracting from the experience of the environment.

After a tour of the park, one might feel their perspective begin reaching back to hold events millions of years prior in mind. Standing once again at the timeline, the imagination projects the end facing the future far into the distance, eventually bending and responding to the circular nature of both Algonquian cosmology and our physical world.

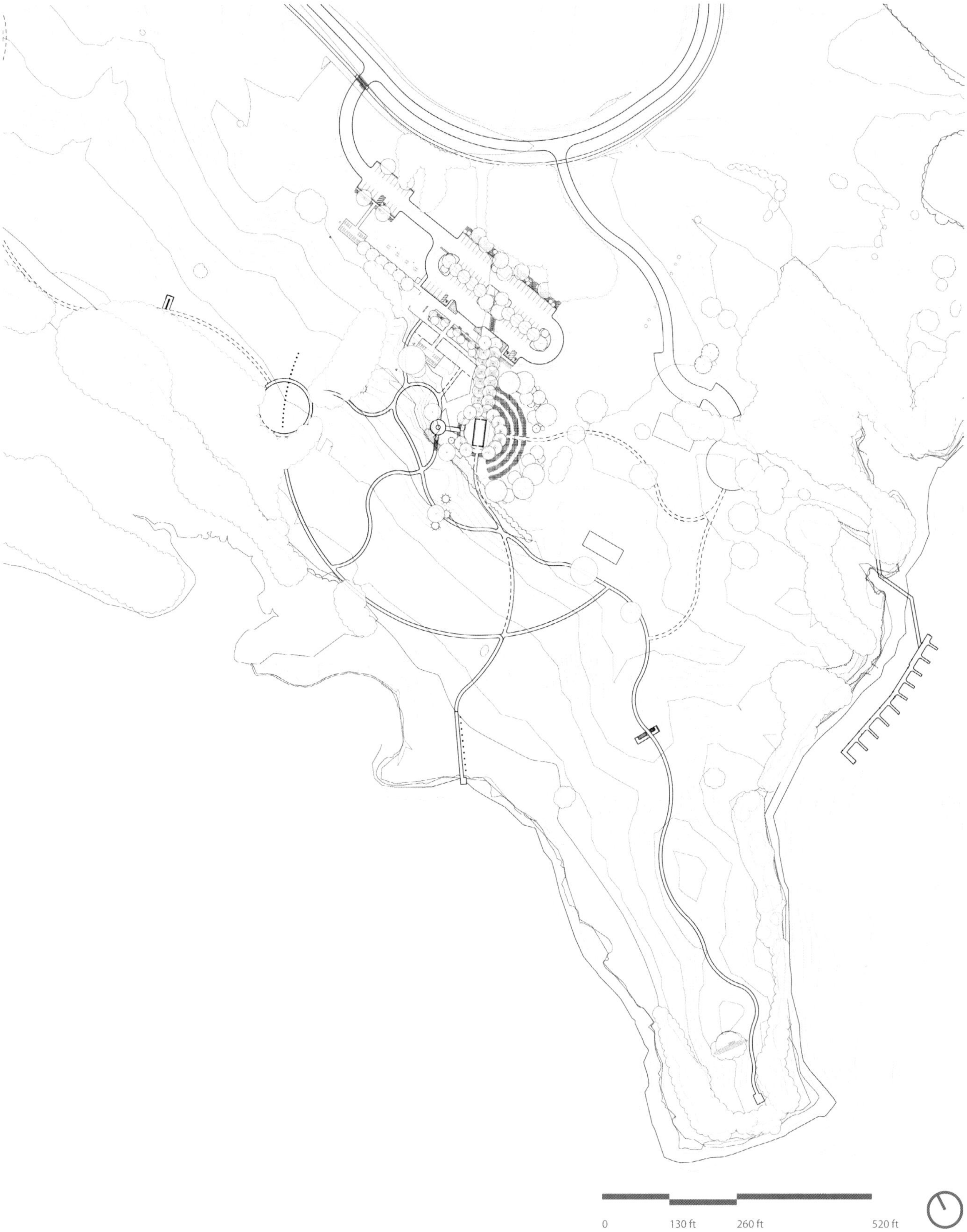

↖ ↑ The park's Tidewater context, with its dynamic wetland ecologies, is compelling for its proximity to Werowocomoco, the historic seat of governance of the Virginia Algonquian Tribes. Machicomoco is the first state park in Virginia to recognize and interpret Indigenous landscapes.

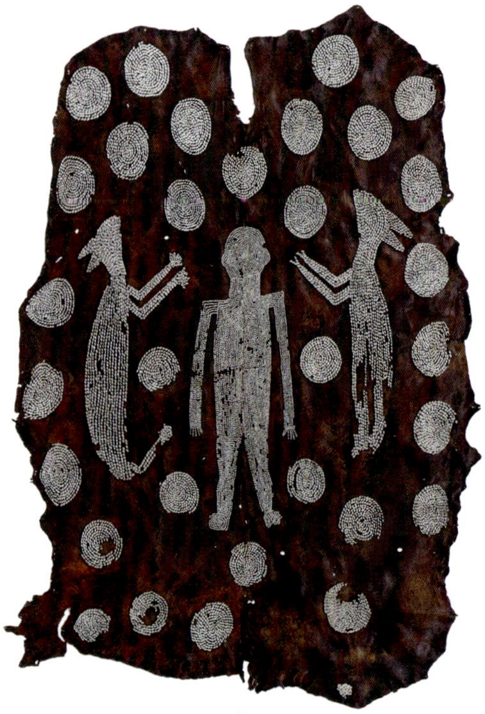

Mantle ascribed to Powhatan (tanned buckskin, shell embroidery.)

Hondius map (after John Smith) of Virginia colony and Chesapeake Bay region, 1630.

Captain John Smith's capture and delivery to Powhatan, paramount chief of Tsenacomoco, engraving by William Hole, published 1624.

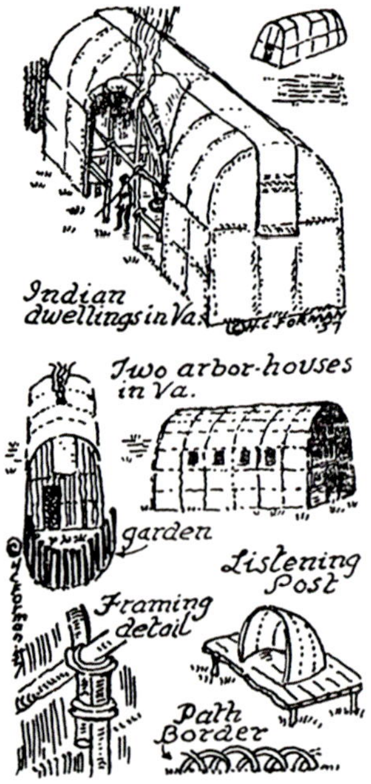

Indian arbor houses illustrated by Henry Chandlee Forman in *Virginia Architecture in the Seventeenth Century*, 1957.

Aerial image of the site of Machicomoco State Park, Google Earth, 2017.

Archaeologists working at the site of Werowocomoco, home of Powhatan, *The Powhatan Landscape*, 2016.

A Pamunkey Indian Reservation delegation presents a large four-point buck to Governor Byrd on steps of the Virginia Capitol in Richmond, c. 1926–30.

↑ The designs of interpretive elements were informed by extensive engagement with leaders and members of Algonquian tribes in the region, as well as with historians and other scholars, to embody the tribes' cosmology and the depth of its roots in the region developed over an estimated 15,000 years. →→ The Pavilion orients visitors with visual narratives of Algonquian history in the region and opens a portal to the wider landscape.

THE ALGONQUIAN LANDSCAPE

← ↑ The design of the Pavilion's open structure is informed by the materials and forms of Algonquian longhouses, which are typically constructed with a frame of bent black locust saplings supporting a shell of woven reeds or bark.

↑ The presence of Algonquian language, symbols, and materials attune visitors to the Indigenous experience in this landscape, to the evolving settlement patterns and provisioning over millennia, to tribal authority and power built and then lost to European settlers, and to the latter-day reassertion of native sovereignty.

↑↑ A timeline embedded in the ground draws visitors into the Pavilion, where graphic panels explain and interpret the site's history ↑ A threshold onto a path is informative, offers seating, and serves as a wayfinding device.

← A woodland path leads to a pier at the river edge, allowing people to pause and observe the activity of the forest. ↑ A park guide leads a group of visiting students through the park and its interpretive elements. →→ Thematic iconography denotes specific tribal meanings at various locations throughout the park, as at the shoreline environment surrounding a pier.

ANGKOR WAT
TIKAL
TROY

↖ The park creates learning opportunities about Algonquian life in the Tidewater region on a shoreline that appears much as it might have when tribes were its exclusive inhabitants. ↑↑ A student reads the timeline of Algonquian history embedded in a walkway. This living timeline leaves ample room for recording future events in the continuing tribal legacy. ↑ Student visitors can observe Algonquian canoe-building methods, among which are scraping out wood with shells and using fire to degrade the interior tissue.

RETURNS TO VIEW

PROJECT
LOCATION

GEORGIA INSTITUTE OF TECHNOLOGY ECOCOMMONS
ATLANTA, GEORGIA

Many of NBW's projects begin with knowledge in mind or at least the strong sense of past ecologies or cultural eras lost or obscured that, along with spatial and functional aims, must be brought to the conscience of the present. For the design of the EcoCommons at the Georgia Institute of Technology, it was intuitive that this landscape could incubate the biodiversity of the Georgia Piedmont ecoregion that metropolitan Atlanta had wiped out. (Poignantly, just sixty-five acres of intact Piedmont forest survive in this sprawl, found at the Fernbank Museum of Natural History near Emory University.) The entire rationale of prioritizing research before design is for the firm to know what it doesn't yet know. For the EcoCommons project, the research stage produced a significant historical finding that raised the scheme to the landmark level.

A collaboration with Georgia Tech over more than twelve years has allowed the firm to guide the school systematically through the phases of a 2004 master plan update for the 400-acre campus. The plan has helped Georgia Tech approach campus projects holistically and avoid random interventions. The EcoCommons comes under one of two major sector plans by the firm to develop a contiguous 100-acre area where landscape intentions take precedence over the placement of buildings. Favorably to its landscapes, Georgia Tech is aiming to create regenerative environments across the campus that will manage stormwater, optimize resource use, and create opportunities for environmental research, all of which position the school as a leader in campus sustainability.

The EcoCommons is part of a forty-five-acre sector around the Engineered BioSystems Building. Its seven acres had previously held parking lots and a campus police station on land where a creek lies buried beneath more than twenty-five feet of fill. A few important large trees survived those surroundings to become part of the EcoCommons, which will be about three-quarters shaded by canopy. The project regraded the site to favor the natural movement of water around the corridor of the buried creek. From the south, at the edge of Ferst Drive, a gentle slope falls toward building fronts lined up along a path to the north. One steeper spot has three playground slides for moving from one path to the path below it.

The paths wander and curl through wild depths of unbridled vegetation where another clear side can't be seen. Most of the landscape is turned over to a diverse range of plant communities. Wet meadow reaches across the middle, following the flow of the buried creek beneath trees that thrive in moist conditions, such as tulip poplar, bald cypress, river birch, and sycamore. Just upslope from the meadow is a mesic mixture of serviceberry, ironwood, and deciduous magnolias; below is mesic forest planted with buckeye, beech, tulip poplar, and river birch. Mixed stands of oak and pine fill the western edge, and the northern edge holds oaks, pines, and hickory trees. Near the center is a clear island of open prairie ringed on the south by a grove of black gum trees where people can relax in any of about a dozen hammocks scattered throughout the space.

A few areas of paved pathways are interrupted by wood boardwalks over a constructed streambed that traces the buried creek's run and directs stormwater downslope, where it is captured for reuse. The EcoCommons catches and stores 50 percent of the stormwater on the site. It has a capacity to filter and hold 500,000 gallons in cisterns for irrigation, sparing that volume from the city's sewer system

and averting the release of municipal water for landscape maintenance. The major catchment lies near the Learning Deck, a long boardwalk with seating and a shade structure designed as an outdoor classroom. Built into the design are capabilities for students and faculty to monitor the vital signs of the EcoCommons environment. Sensors on the site track air quality, temperature, and humidity; soil moisture and temperature; and water volume and quality. Students can sample soil to track its organic and microbial content, respiration, and drainage rate. The data and trends bring immediacy to environmental education, allowing for up-close observation and training and giving students the ability to design enterprising research programs.

The firm's own early investigations of the site involved following paper trails into the past through property records, among other documentation. In 2009, the university demolished a building that stood on what is now the west side of the EcoCommons. Before the university acquired the property, the building was The Pickrick restaurant, owned by Lester Maddox, a twice-failed candidate to become mayor of Atlanta who went on to become governor of Georgia from 1967 to 1971.

Maddox was a segregationist. On July 3, 1964, a day after the passage of the Civil Rights Act of 1964, three Black seminary students, George Willis Jr., Woodrow Lewis, and Albert L. Dunn, attempted to stage a sit-in to desegregate The Pickrick. Maddox, who was armed, and patrons wielding pick handles forced the students off the property. The students sued Maddox under Title II of the Civil Rights Act, which covers public accommodations. They were represented by the civil-rights attorney Constance Baker Motley (who in 1966 became the first Black woman to serve as a federal judge). Their case was joined to a separate case—a federal complaint opposing the Civil Rights Act, filed by a co-owner of an Atlanta hotel—and was heard by a federal court in Atlanta, which upheld the new law and told both business owners it required them to serve Black patrons. The hotel owner appealed alone to the Supreme Court, which upheld the Civil Rights Act unanimously. Maddox refused to integrate his business, and he closed The Pickrick in early 1965.

The firm reserved the area inside The Pickrick's footprint as a site to commemorate these events. A walkway runs along a concrete wall that encloses the front of the site to represent the barrier of segregation. The prairie, to the south, encroaches on the walkway to pinch it more tightly where three openings break the wall to admit entry into the ghost of the restaurant's interior. Inside, three concrete pylons stand higher than the barrier wall to represent the students and the paths they opened for racial progress. In the middle distance, beyond a low wall of weathered steel, three longleaf pines are rising from a rough expanse of meadow, projected as living legacies of the students.

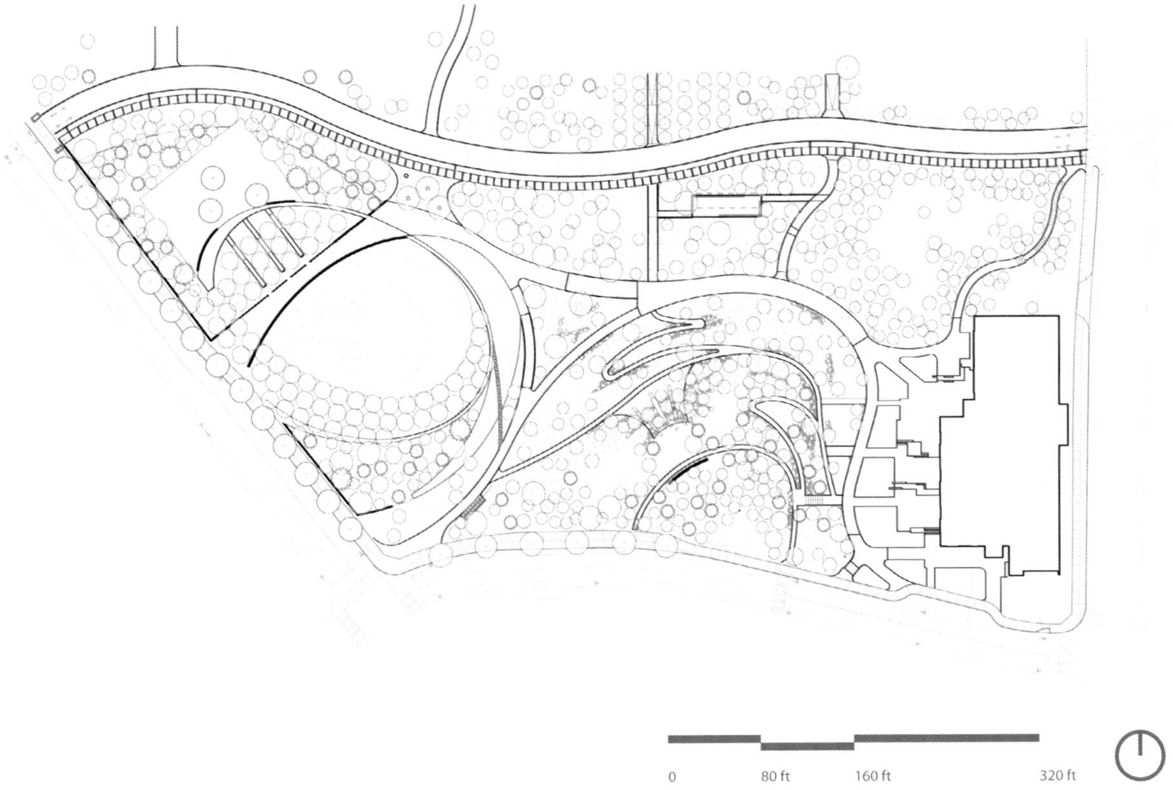

↖ The campus landscape opens opportunities for respite as well as ecological research. ↑ The circulation pattern of the EcoCommons site inserts several kinds of universally accessible paths through the interior, at many points creating moments of total immersion in a landscape layered with Piedmont plant communities.

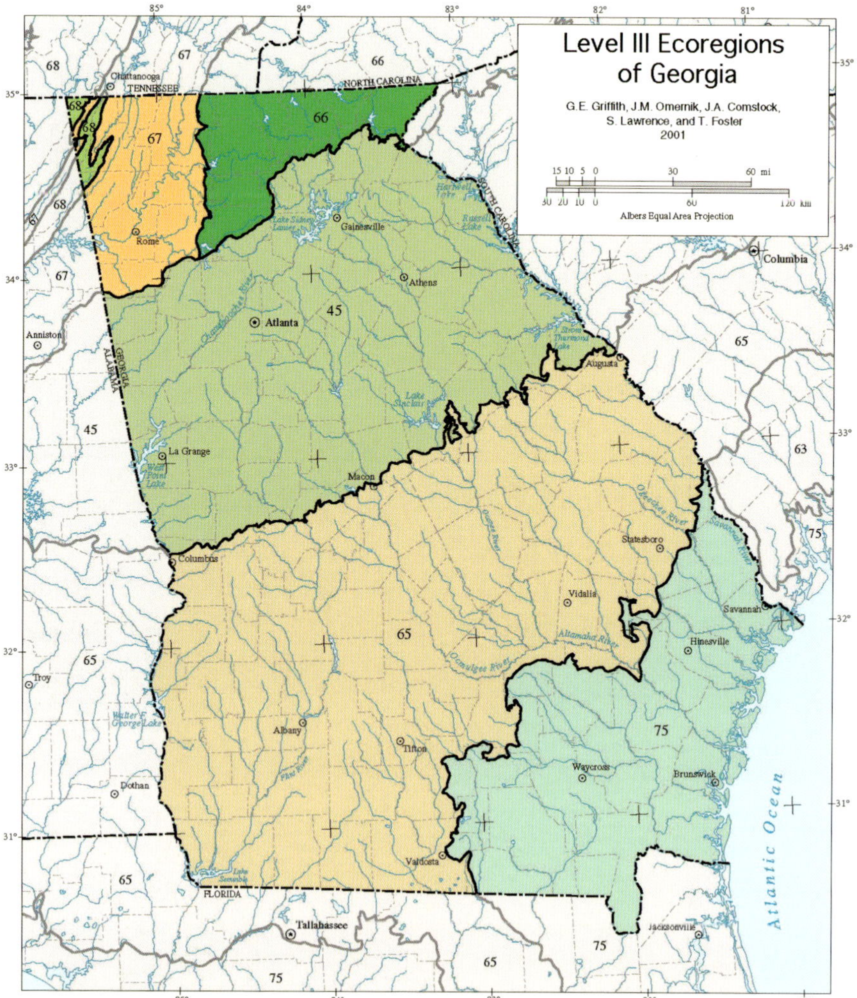

Atlanta sits within the Piedmont ecoregion of Georgia.

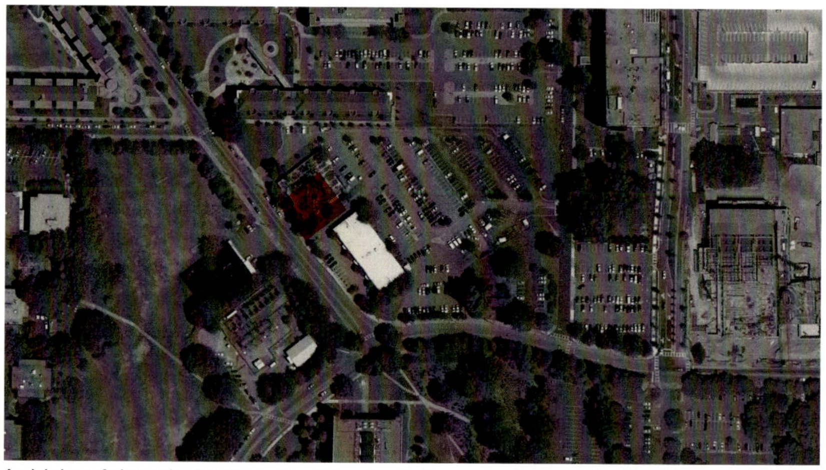

Aerial view of site and extensive parking lot surface before redevelopment; The Pickrick restaurant in red, 2007.

Projected succession of site vegetation, carbon sequestration, water filtration, and avoided runoff.

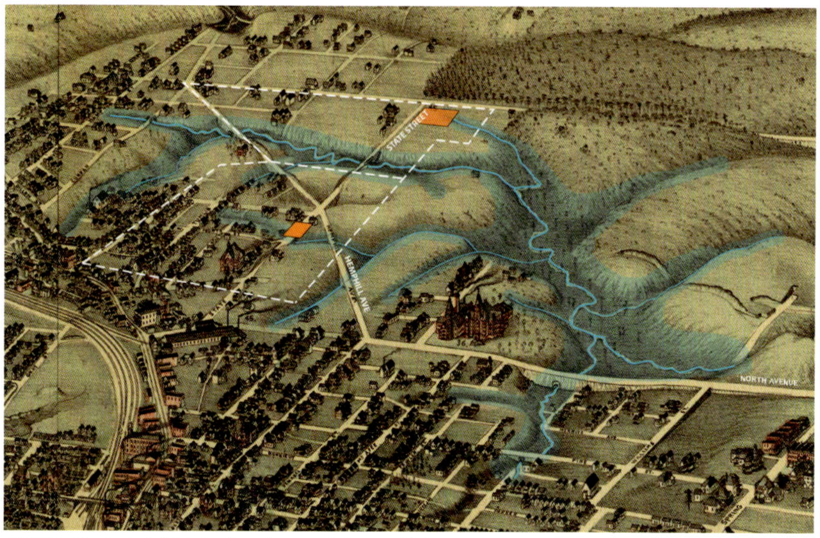

A bird's-eye view of Atlanta, Saunders and Kline, c. 1892.

Lyndon Johnson signing Civil Rights Act, July 1964.

Crowd forms to prevent African-American seminary students from entering The Pickrick restaurant, 1964.

The Pickrick restaurant on the future EcoCommons site.

Constance Baker Motley, a future federal judge, was co-counsel for The Pickrick plaintiffs.

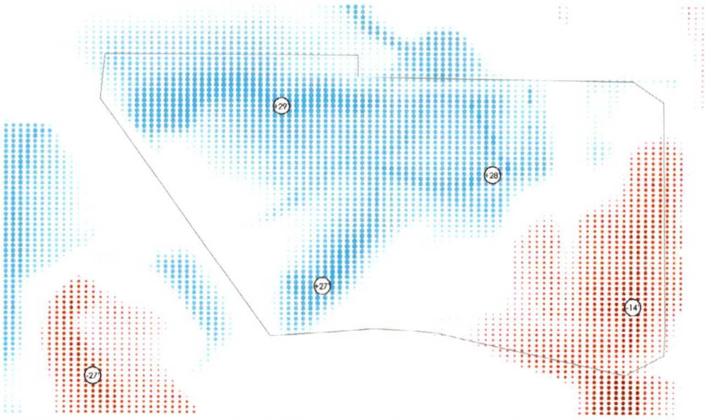

During on-site excavations, buried trees were discovered as far as twenty feet below the surface as evidence of a much lower historic topography that had been filled.

↑ The project began on a paved site over a stream channel buried under earthen fill. The design created an analogous watercourse that directs drainage downslope to a constructed wetland around an outdoor classroom deck. Wood-plank sections of pathways indicate their passage over the watercourse.

←← ↑↑ A contemplative space set within the site of the former Pickrick restaurant pays tribute to three students who attempted to defy segregation in 1964 and were threatened by a mob. ↑ →→ A boundary wall represents racial barriers, with openings, signifying resistance, into an austere garden of black gum trees. → Wood benches invite people to pause in reflection.

↑↑ A learning deck sits above a constructed wetland area that accepts drainage from upland areas of the site.
↑ The planting strategy inserts a characteristic native Georgia Piedmont ecosystem that people can engage in multiple ways.

↑↑ Various areas of the commons are carved out as places to enjoy quiet and solitude amid the dense and diverse native plant communities. ↑ A hammock grove offers a place for students to relax beneath a grove of black gum trees that turn crimson in fall.

IV

ESSAY

182 THE SAVING POWER
Robert Pogue Harrison

PROJECTS

188 FLIGHT 93 NATIONAL MEMORIAL
Shanksville, Pennsylvania

202 AGA KHAN GARDEN
Edmonton, Alberta

218 MEMORIAL PARK
Houston, Texas

ESSAY

THE SAVING POWER

Robert Pogue Harrison

A week before sitting down to write this essay I flew to Houston for an extended walking tour of the sites that NBW took the lead in designing inside the city's immense Memorial Park. After a couple of hours, something surprising took place: I felt an obdurate metaphysical gloom inside of me quietly lift for the first time in years. It wasn't just an aesthetic response to the vitality that NBW had reinfused in the sites. It was the realization of what can be accomplished when a visionary landscape architecture firm is backed by the will and resources of various municipal, corporate, private, and public organizations. I had not fully acknowledged until then of how much I had succumbed to a psychic condition shared by many other people around the world today, perhaps even a majority of us, whether we're aware of it or not. I will call it eco-depression, or better, Anthropocenic depression, a pessimism over the ways humanity's accelerating subjugation of our planet's zone of life are transforming habitats into ecological and spiritual wastelands. The pessimism mutates into depression when one starts believing that only humanity's elimination could put an end to the degradation of our planet's biosphere. My visit to Memorial Park disabused me of this assumption. I saw evidence that, under the right conditions, humankind's conjugal relation to nature can regenerate from within the Anthropocene rather than beyond it.

At the heart of a city without zoning laws, the same buildout machine that had turned the forests, marshes, and prairies of the Gulf Coastal Plains into the monstrous sprawl of Greater Houston had teamed up with NBW to rescue one of those sites from its past malpractice by revivifying and, as well, re-historicizing it. The technologies that had bulldozed and paved the areas around the Eastern Glades in the past were used to dig up and remove its roads and parking lots. Here a new urban gateway comes into being like a promise of happiness (*une promesse de bonheur*, Stendhal's definition of beauty), and the site's secular history (centered around Camp Logan) once again comes to light. Judiciously seeded native species spring up with injudicious exuberance. Pedestrian walkways recede into woodlands. A sense of enclosed infinity pervades the site. If there is such a thing as the world soul, or *anima mundi*, as the ancients called it, NBW succeeded in resurrecting it in Memorial Park.

I have seen my share of private oases, magnificent gardens, national parks, and heterotopic sanctuaries protected from the encroachments of human civilization, yet NBW's reclamation of the land for urban and public purposes in Memorial Park offers a blueprint for what a regenerate relation between city and nature can—and should—look like in the Anthropocene future. It looks to me like a landscape that has welcomed humanity's presence as a friend or a guest, almost as if the firm had reintroduced into these sites the laws of hospitality between land and citizens that have been debased by decades of indiscriminate urban development. The hospitality goes both ways, for nature is invited into the city as well. This mutual hosting between nature and civic space puts both parties on their best behavior.

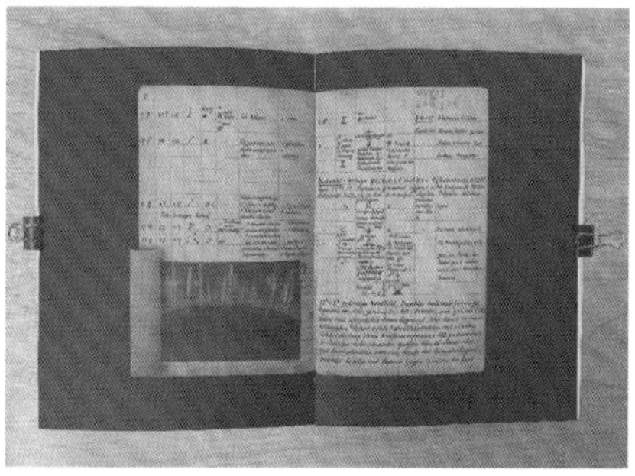

fig. 1 Illustrations from *A Shelter for Bells*, a book of writings by Hans Jürgen von der Wense.

The German poet Friedrich Hölderlin once wrote: "Where the danger is, there the saving power also grows." During my Houston visit, I felt the upsurge of a saving power at the heart of the danger. How now? Why there and not elsewhere? Don't we see the saving power at work in our conscientious transition to electric vehicles, heat pumps, electric cooktops, and wind power; in our ingenious proposals for carbon reduction, removal, and storing; and in any number of other well-intentioned green initiatives? Alas, there is more to the saving power than clean energy, eco-friendliness, and countermeasures. The saving power is not a corrective. It is not a resource. It can be summoned but not extracted. Its alchemical reaction to danger does not lend itself to calculation, manipulation, or problem-solving. We humans do not wield the saving power, nor do we harness it. At most we can release it.

When it comes to landscape architecture's role in this release, what is required foremost is a special kind of clairvoyance, one that sees through a site's urban accretions to the geology, ecology, and regional history that underlie that site. Every place has its "underland," as it were, and it is there—beneath the superstructures of our built environments—that the cultural as well as natural elements of place cumulate their energies. A vision of the underland does not by itself release the saving power. Vision must be followed by a judicious art of removal—the removal of whatever smothers and suppresses the site itself. Removal prepares the way for revealing in a landscape. It throws into relief all the more distinctly what the removal process discloses.

NBW offers a paradigm for this revelatory approach to landscape design. In many ways that approach reminds me of Michelangelo's art of sculpture, which, if we believe what Michelangelo himself claimed, consisted in a diligent removal or clearing away of the excess matter in a block of marble until the form hidden within the stone was exposed. In the master's words: "Every block of stone has a statue inside it, and it is the task of the sculptor to discover it by dint of removal (*per forza di levare*)." We could call this method "abstraction," in the etymological sense of abstrahere, to draw away or remove. There is a difference, to be sure. Where Michelangelo "saw the angel in the marble and carved until I set it free," Thomas Woltz looks not for angels or disembodied forms but rather for the intricately layered natural and cultural nexus underlying a given site. Only after that nexus is envisioned and reconceived can the work of removal begin. Once the superfluous accretions have been removed and the essential revealed, the site's enduring substrate provides the measure for what the landscape architect may now add by way of design. In Woltz's words: "We begin by looking at the whole site—the large-scale plant communities, the geology, the solar angle, the hydrology, the history—and all that research helps build a narrative that becomes the framework for our design." The narrative, in other words, lies embedded in the site, and it is brought to view by a discerning, holistic looking. To formulate the process in an overly schematic fashion: vision is followed by removal, revelation is followed by design.

The procedure has something in common with Phenomenology's method of "reducing" an entity to its most essential features (to its

fig. 2 Hans Jürgen von der Wense drawings depicting wind types.

"eidetic" features, in Phenomenology's technical jargon). By eliminating the inessential, the phenomenological reduction allows the subject of apperception to "intend" the phenomenon more fully, where "intend" means to grasp it in its full self-manifestation. If we think of place as a phenomenon—and there is every reason to do so—we could say that the Anthropocene in recent times has militated against that plenitude by paving over, disfiguring, smothering, or obscuring the basic lineaments and features of place. Where a site's natural and cultural history is covered over, it fades away from the realm of appearance and intelligibility. In its highest vocation, landscape architecture reverses this disappearing act through its art of removal and laying bare, restores the phenomenological plenitude of place. In Woltz's words: "The urgent problem today is to reveal place and connect people to place." Revelation enables connection.

By rendering a site more fully present to human apperception and engagement, the architectural act of revealing recalibrates the relation between people and landscape in such a way that a resonance or attunement between the two becomes possible. Attunement reaches beyond ocular vision into our psychic, kinesthetic, poly-sensuous experience of place. Here we enter into the more elusive aspects of landscape design, namely, its power to respond to the underlying tone or "melody" of a place. In a striking three-page essay called "The Psychology of Place," the author and outdoorsman Algernon Blackwood declares that places have their musical signatures. He speaks specifically of the note of a place: "Whereas most of us are quick to get the 'note' of a person, for like or dislike, it has been my experience that only a few respond vividly to the 'note' of a place" (*The Face of the Earth*, 154). The German writer Hans Jürgen von der Wense ❶ ❷, also an inveterate outdoorsman, declares something similar if not identical: "People see so much, but they do not listen to nature . . . Each thing on earth has its inner melody . . . I merely excavate the tone of all things" (*A Shelter for Bells*, 56, 144, 90).

From what I not only saw but also heard at Memorial Park, I would say that Thomas Woltz has a special ear for a landscape's note or inner melody, and that his designs heed it in ways that allow place and person, or mind and phenomenon, to reverberate with one another. "Mind" here should be understood in the expansive sense of human apperception of, and attunement to, the *Umwelt*, or surrounding world. To heed the note of a place entails a great deal more than what reaches the cochlea of the human ear or registers on our recording devices. Heeding involves feeling, sensing, intuiting, and responding to the geo-sonority of place, by which I mean the insensate sound of the earth's priority, its antecedence to human history, and its transcendence of a punctual present. I believe that humans, like other life forms, have—or at least used to have—an innate kinesthetic attunement to the geo-sonic frequencies of the earth on which we tread, and that inspired landscape architecture lends an ear to such attunements.

I would not go so far as to say that every place has a *genius loci*, or presiding spirit, but many once did, and under the right conditions the genii will return from their scattered exile. Where a spirit comes out of hiding and re-occupies a place, it manifests as a subtle vibration of presence, a quivering of the surrounding stillness (I'm thinking here of the Eastern Glades in Memorial Park). Landscape designs at their most inspired become instruments for euphonic resonance of these vibrations. Thus, the release of the saving power revitalizes the land but also the "call" of the phenomenon—a call emanating from the site and heeded by the heedful mind. Such mindful heeding occurs not somewhere inside our heads but in the elusive correspondences between human sentience (in the expansive sense) and the recessive depths of what surrounds and envelops us in a given place.

If hell is the impossibility of reason, as a sage once said, it is also the impossibility of hearing. The Anthropocene's unworlding of the world brings with it the deafening noise of vendors, vatics, and vituperators; the strained accents of self-assertion and self-mutilation; the unrelenting drone of our Borg collective; and the chatter of inauthenticity. To preserve a margin of sanity we need places in our midst where a certain kind of listening can take place. Where we can no longer hear the call of the phenomenon, or the "note" of a place, we become detached from the context of permanence that measures our mortal, transitory sojourn on the earth. As Hannah Arendt put it, referring to our modernity: "The world becomes inhuman, inhospitable to human needs—which are the needs of mortals—when it is violently wrenched into a movement in which there is no longer any permanence."

During my visit to the Eastern Glades in Houston, amid the site's pathways, trees, waters, and blooming native species, I could hear a tone or hum that sounded the underlying permanence of place. My distinct impression was that everyone else wandering in the site could hear it too. This geo-sonority was altogether synesthetic in nature, like Baudelaire's forests of symbols, where perfumes, sounds, and colors mingle in the distance "like prolonged echoes in a deep and tenebrous unity." When landscape design finds ways to release the saving power, this expansion *des choses infinies*, as Baudelaire called it, takes place in the very midst of our pulsating and pullulating urban environments. On a planet rapidly shrinking all around us, we need to make as much room as we can for such local expansions of infinity, and that means we need more landscape architecture firms like Nelson Byrd Woltz.

THE HEART LEADS WHERE THE MIND CANNOT

PROJECT
LOCATION

FLIGHT 93 NATIONAL MEMORIAL
SHANKSVILLE, PENNSYLVANIA

The Flight 93 National Memorial honors the forty people who died on September 11, 2001, aboard a United Airlines aircraft, having thwarted a plan by hijackers to crash the plane into the United States Capitol. Like the memorials at the other sites attacked on 9/11, in the World Trade Center in New York and at the Pentagon in Arlington, Virginia, the Flight 93 Memorial has in its design a sense of its own specificity in shaping the public, shared experience of honoring the victims of the attacks. It binds a design of remembrance not merely to the location of the tragedy but to the place, to its precise coordinates but also to the character of what lies on and around it.

The difference of the Flight 93 Memorial is that the attack concluded, at 10:03 a.m., somewhere other than its intended target, a fact that demonstrates the heroism of passengers and crew. That difference also inescapably evokes the existential randomness of the events that day. The design of the memorial to Flight 93, by NBW with Paul Murdoch Architects, captures the solemnity needed as the afterword to this loss and meets the abstract, universal questions to which no person expects answers.

The design was chosen by a jury of relatives of those who died, design professionals, and public officials in a two-stage international competition managed by the National Park Service. The jury appreciated the NBW design's quiet focus on the tragedy and its human costs as well as its spatial imagination in the setting and responsiveness to seasonal change in a remote rural environment. An enormous circular meadow spreads out on a slope above the crash site. The jury commended the "elegance of the void" in representing losses for which words fail.

The memorial remains in the care of the park service. Its boundaries encompass 2,200 acres between the entrance on U.S. Highway 30 and the approximately three-mile distance to the crash site. The firm began the design process by studying the nature of the site and its surroundings. As chance, and only chance, would have it, the site where the plane met the ground is a former strip mine, nominally reclaimed by the mining company responsible but with its damage far from reversed. The land had undergone more than fifty years of coal extraction at the surface and below. Before mining began, the memorial grounds held a mix of farms, fields for herds, and forests widely disrupted by logging. The site lies in Somerset County on the Allegheny Plateau, about 2,400 feet above sea level, with a ridge line that runs along much of its length from north to south. The visible surroundings are hilly, with the Laurel Highlands to the west and the Allegheny Mountains to the east.

On the approach road to the heart of the memorial site, long views to the west look off to distant ridges. Not far beyond the entrance, a sculptural chime structure called the Tower of Voices rises ninety-three feet from a large field and in the wind pours the somber tones of forty polished aluminum chimes across it and into a realm without edges. Wildflowers encircle the terrace, and in the field around it, pines are planted in concentric rings across the landscape.

The road ends at the visitor center, a long building of monumentally high concrete walls, broken by a walkway that marks the final path of Flight 93. The building is in the form of an arc that follows a circular path about two miles around a 140-acre meadow called the Field of Honor, across a slope above the point of impact and debris field, which is known as the Sacred Ground. The

Sacred Ground had been greatly altered by mining and during reclamation had been seeded with merchant-grade grasses and clover. It is now planted as a meadow that mixes grasses and forbs suited to the region. The firm found surrounding tracts of land that were relatively undisturbed, with intact forest communities of species that might transfer aptly back to the site. The region originally held Northern Hardwood Forest communities of maple, birch, beech, and white ash, with some hemlock and eastern white pine, and an understory of rambling rose, black raspberry, hay-scented fern, reed-canary grass, and coltsfoot among other species.

Mining had degraded the site's soils, requiring a substantial amendment to support a successful planting regime. There are also the challenging slopes and exposures of the Field of Honor and the Memorial Groves that enclose its edge—forty groves of forty trees each planted in memory of each passenger and crew member on Flight 93. From the top of the ridge to the bottom, the land tilts down an elevation change of about 200 feet from north to south and lies exposed to harsh westerly winds.

Across the sweeping slope of the Field of Honor, the landscape strategy is relatively simple and emphasizes the power of scale as a preamble to the Sacred Ground, which lies in a low, flat area next to a grove of dark-green hemlock trees. Near the top of the Field of Honor an arc of red maples, which turn crimson in fall, follows its edge. The circular path takes time to walk down and around to the groves. Near the bottom, a bridge crosses a pond at the meadow's edge. There is a small visitor pavilion at the lower end of the ring path. Just beyond it to the left is the Sacred Ground, a meadow of high grasses. It is very quiet. The hush deepens the sense of presence at this place. On a July day, with the hemlocks enclosing the background, a deer bent down its head in the middle of the green to browse amid white and yellow wildflowers, alone in the field except for a single large boulder that marks the point of impact, the point of the transformation noted by an early responder to the crash of Flight 93: "A common field one day. A field of honor forever."

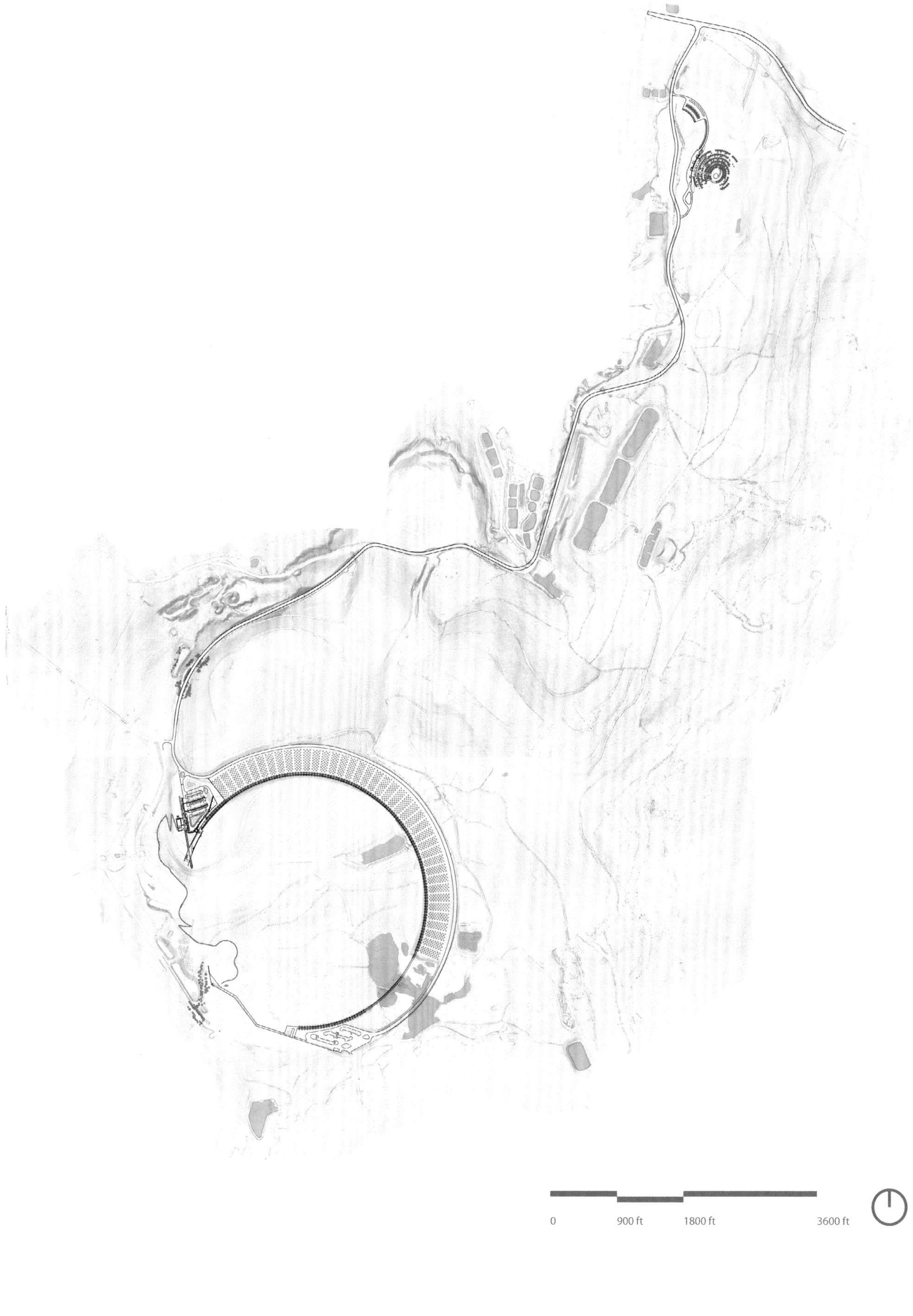

↖ A circular path encloses the Field of Honor above the crash site with an allée of maples. ↑ The procession to the point where Flight 93 crashed leads visitors through rolling terrain along one edge of the 2,200-acre park in the Central Appalachian Upland region.

Wednesday, September 12, 2001 THE PHILADELPHIA INQUIRER A3

A DAY OF TERROR

"I don't think there are enough words in the English language ... to describe the range of emotions that I as a Pennsylvanian feel after witnessing what may be the most brutal act of terrorism in American history."
— Pennsylvania Gov. Tom Ridge

LAURENCE KESTERSON / Inquirer Suburban Staff

Pennsylvania State Trooper Jeff Lenis stands guard over the scene of the plane crash in Stonycreek, Pa. Investigators were surveying the damage and searching the debris yesterday.

Crash in Pa.: 'We are being hijacked!'

Pennsylvania State Trooper guarding the crash site on September 11, 2001.

A temporary memorial quickly grew in the aftermath of the crash.

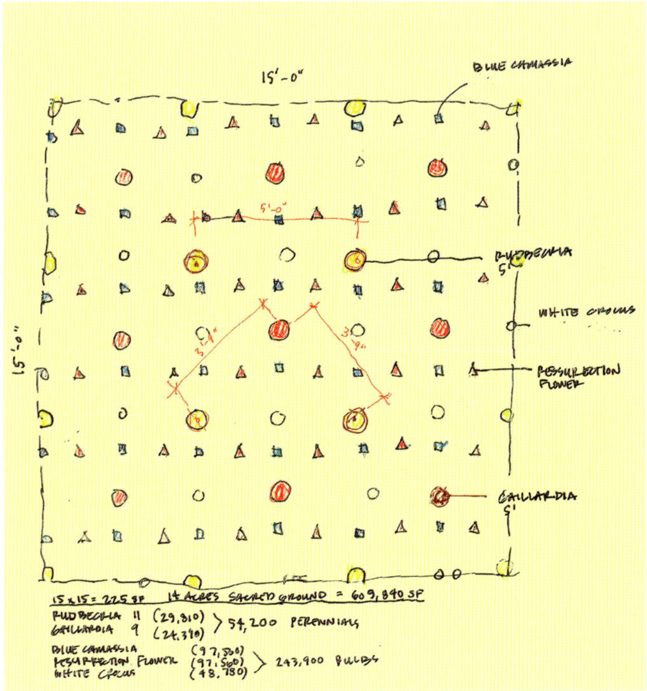

Draft plan for perennials planting across the Sacred Ground.

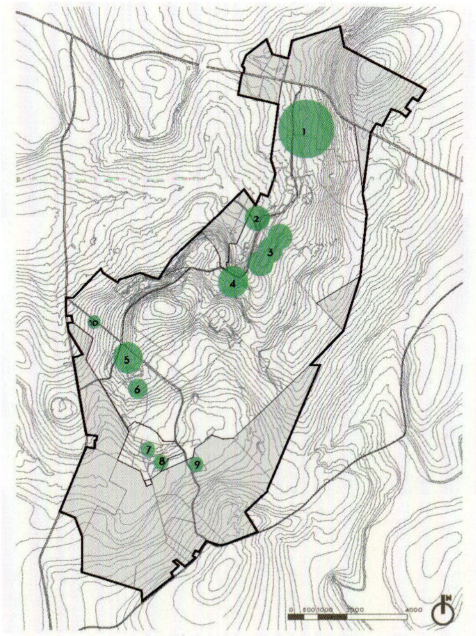

Areas of environmental concern where contaminants from various coal mines and scrap-yards were identified.

Landscape-scale study showing design iterations across the 2,200-acre site.

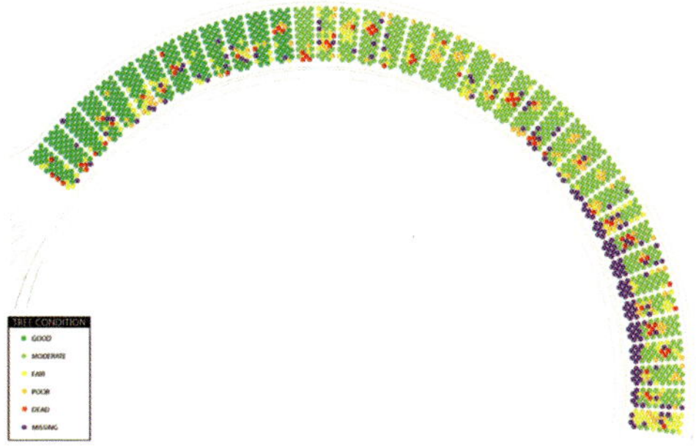

Tree conditions as documented among the Memorial Groves, 2020.

Surface mine operations of PBS Coals on the site, which ended in 1995.

Dragline that previously mined for coal seams flies an American flag to show support to those who worked at the crash site.

↑↑ A paved walk through an opening in a visitor center wall traces the path of Flight 93. ↑ Down the slope, a boulder marks the point of impact. The field immediately surrounding that point is reserved as a sacred site accessible to immediate survivors of the victims. →→ At the heart of the memorial is the Field of Honor, ringed by an allée of red maples and the forty Memorial Groves, the visitor center, the Memorial Plaza and Wall of Names, the visitor pavilion, and the Sacred Ground surrounding the crash site.

↖ The Memorial Plaza is bounded by a canted wall. ← A folded, marble Wall of Names honors the victims of the Flight 93 hijacking. ↑ Two large geometries intersect and frame views of the flight path and the Field of Honor. The meadow stands in stark contrast to the monumental scale of the architecture. →→ A footbridge completes the two-mile circular path below the Field of Honor.

↖ Forty groves of forty trees each—sugar maple, white oak, and elm—form a living memorial to each passenger and crew member. ←↑ The sugar maple and red maple varieties were specified for their turning a vivid red around the area's first frost date in early September.

BETWEEN HERE AND PARADISE

PROJECT
LOCATION

AGA KHAN GARDEN
EDMONTON, ALBERTA

The Gift

The Aga Khan Garden at the University of Alberta Botanic Garden in Edmonton is a gift to the university from His Highness the Aga Khan, the Imam of the Shia Ismaili Muslims worldwide, a gesture of gratitude for the long bond between the Ismaili people and Canada upon its 150th year of nationhood. It also marked the 40th year of a collaboration between the university and Aga Khan Development Network. The network, known as AKDN, and the affiliated Aga Khan Trust for Culture work in communities around the world to improve people's lives through education, health, entrepreneurism, crisis resilience, the arts, and the built environment. The gift of a garden shares a place essential to Muslim life, a realm of pure transcendence "where the ephemeral meets the eternal," in the words of His Highness. It also resonates from the core of AKDN's conviction that nature lifts the well-being of people by offering joy and respite from urban confusion. This garden joins more than a dozen other public landscapes that AKDN has built or conserved in Cairo, Dushanbe, Bamako, and elsewhere. The Alberta garden is the northernmost garden drawn from Islamic traditions in the world, a condition that called for imaginative interpretations by NBW to suit the cold climate of the Eastern Alberta Plains.

Pattern and Form

Research for the design involved visits to several notable Islamic gardens in India and Egypt. Mughal gardens emerged as a major influence, particularly the Rham Bagh outside Delhi, built in 1528. These gardens arose in agriculture and evolved into expressions of Paradise; the Qur'an makes numerous allusions to gardens as exalted sanctuaries of purity. The firm studied with artists in the Moroccan city of Fez in generating the types of geometric patterns, grounded in numerology and rendered with precise symmetries and proportions, inherent to Islamic arts and design. Pattern repetition implies the universal order of existence and invites contemplation at the threshold of earthly nature and the promises of Paradise. The entry into the Aga Khan Garden signifies such a transition between realms. A floating walkway leads down gently through a woodland garden, or *bagh*, and twice turns 90 degrees as it leaves the ordinary for the ineffable. A path rises to the *talar*, a stone pavilion of twenty monumental pylons that looks across the expanse of the *chahar bagh*, a four-part garden suffused with water symbolic of the four rivers of Paradise, set within wide, patterned walkways. Among the garden's long walls and its stairs, form has primacy; the detailing is austere, constrained principally to subtle reveals that underscore copings. Patterns across horizontal surfaces allude to the earth, whereas on vertical elements they relate to the heavens. Integral patterns appear across gateways and screens—dodecagons interlace along the rail panels of the *talar*, and radial star motifs are stenciled on a weathered steel screen that defines a *jilau khana*, or forecourt, and they also form the body of a stone balustrade enclosing the *mahtabi*, a viewing platform between a water garden and the Calla Pond, a reconstructed wetland. The visual reticence of structural elements reserves consciousness for meditating on the whole of the garden and its communion with the light of the sky.

↑ The *talar*, traditionally a throne hall, is made of tensile fabric stretched between tall limestone piers, a contemporary interpretation of an inhabited ruin overlooking the core of the garden. The concept was inspired by the Emperor Akbar's city, Fatehpur Sikri, abandoned in 1610 owing to water scarcity.

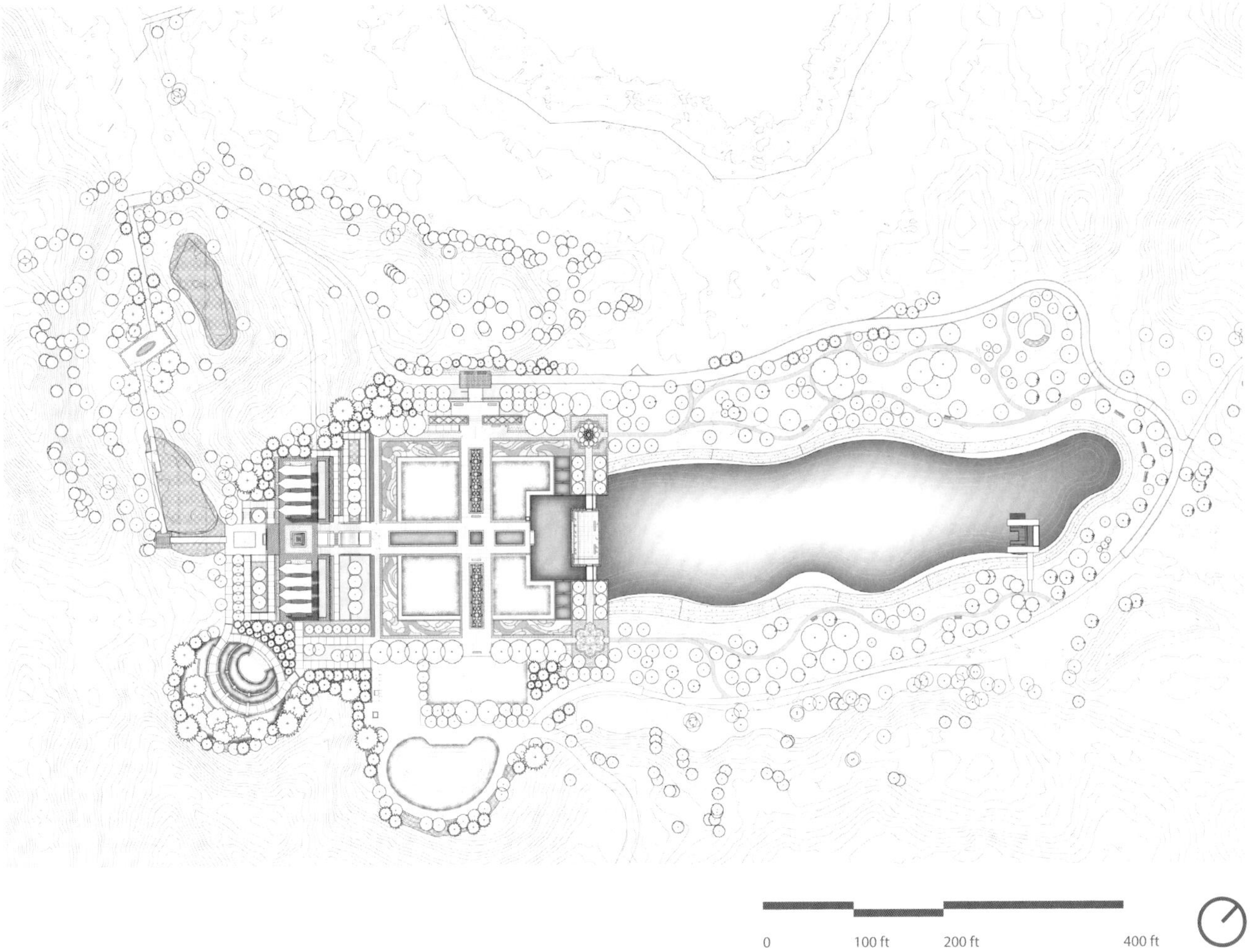

Water

Water in Islamic gardens is divine, an eternal gift from a generous God. The Qur'an glorifies appearances of precious wellsprings, streams, and flowing fountains of water—sweet, nourishing, and abundant. Water entered as lifeblood through channels built into early agrarian gardens in arid lands. It evolved as the element essential to the absorbing sensory experience of a garden with its pleasurable smell, the music of its movement, and the sight of its reflections in the light. On the path into the woodland *bagh* of the Aga Khan Garden, the leaves and limbs of the forest overhead glow in reflections of a quiet pool in a long oval basin of black stone set in a clearing, a moment that untroubles the mind for the journey toward the garden's watery generosity. The path stretches down through the forest to a low stone fountain that spills from a wall scupper into a basin. The final turn takes the path to a wall inset with a waterfall of fine jets that spills downward as a woven diamond pattern into a small pool; its sound resonates against the wall and rises to fill the air of a terrace below the *talar*. Its sound recedes on the progression to the plinth of the *talar*, where a wash of sound issues from a cascade of fountains that begins at a low black stone cube set in a pool and slips out of sight to re-emerge as a wide sheet of water sweeping over a niched wall, the *chini khana*, drawn from Mughal designs, and the *chadar*, a "shawl" of water coursing over a stone slab with fine patterns carved onto its face. The water travels forth into the channels of the *chahar bagh*, ultimately reaching the richly planted wetland around the Calla Pond.

Seeds

Alberta's landscapes require lifetimes worth of repair. Across Range Road 263 from the main entrance to the botanic garden are several of the 157,000 well heads for extracting oil or gas in Alberta, which also had fires across 5.4 million acres of its land in 2023. The Aga Khan Garden tangibly expresses Islam's regard for nature as a creation of God that requires human stewardship. The design posits the garden as a salve to recuperate the pervasive wounds of the province. Traditional gardens at lower, warmer latitudes hold sensuous plantings in profusion for their fragrance, beauty, and fertility. For the frigid, dry climate of Edmonton, the firm's research sought surrogates among species native to Alberta to thrive and reveal beauty in punishing cold and, more pointedly, those for which seeds were scarce for the purposes of ecological restoration. A garden where the seeds could proliferate would become an analog of Islam's regenerative spirit. Specialists at the botanic garden helped inform planting choices with data from its evaluations of plant hardiness combined with the site's hydrological profile. This knowledge, along with inspirations from Islamic literature, garden precedents, and Persian arts, converged in a selection of perennials, bulbs, stone fruits, berries, and roses suited to the garden's setting. Beyond the walled precincts of the garden, informal stroll paths lead through areas designated as a *bustan,* or orchard, populated by twenty varieties of fruit trees. Its cultivation calls forth centuries of agrarian and horticultural wisdom invested in the garden arts and projects it toward a future of healing for Canada.

Botanical illustrations from a thirteenth-century Arabic translation of the first-century Greek pharmacopoeia *De Materia Medica* by Dioscorides.

Design precedent: Plinth and canal, Ram Bagh, Agra, India, c. 1526, renovated 1615.

Design precedent: Terrace stair and *chadar* (shawl) patterned cascade, Ram Bagh.

Haftvad and the Worm, folio from the *Shahnameh (Book of Kings)* of Shah Tahmasp, c. 1540.

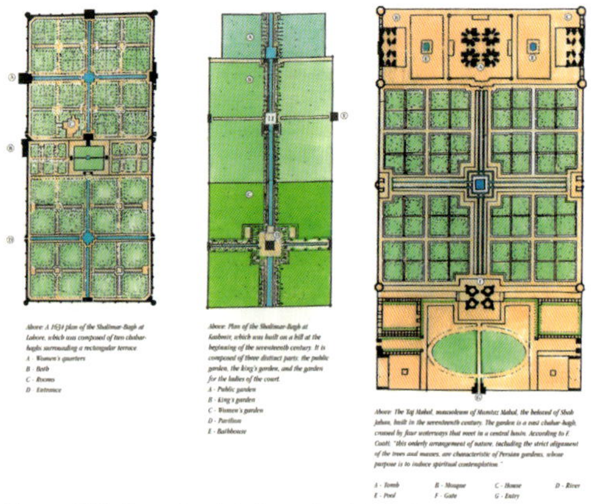

Plans of *Shalimar-Bagh*, Lahore, Punjab, Pakistan, 1634; *Shalimar-Bagh*, Srinagar, Jammu and Kashmir, 1642; *chahar-bagh* of Taj Mahal, Agra, India, 1631–53.

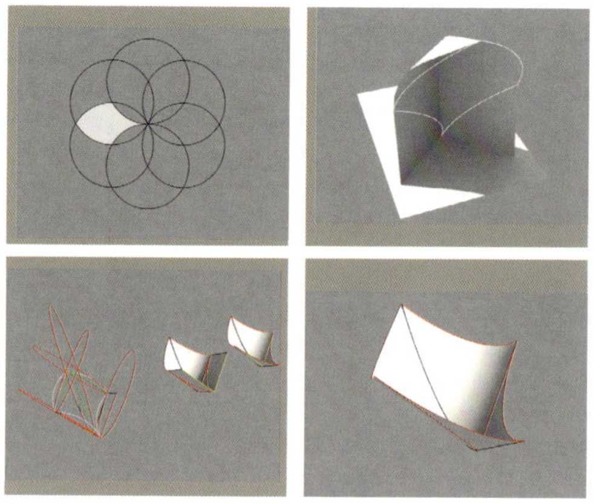

Sequence of pattern derivation of *chini khana*.

Drawing of patterns for panels of *mahtabi*, or garden overlook.

Aerial photograph of oil exploration sites in the Athabasca tar sands region of northeastern Alberta.

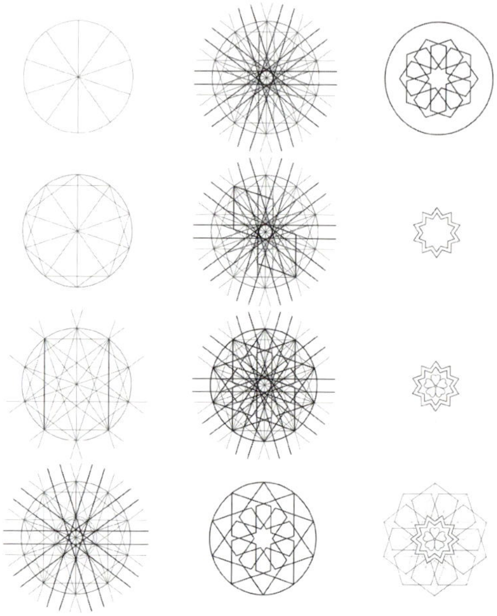

Pattern development study.

Traditional tile carving.

↑↑ The firm's research and practice in the patternmaking integral to Islamic garden design informed the process of creating site elements. ↑ Constructed tanks are incorporated into the system of water courses and reserved for the cultivation of sedges and rushes for use in restoration ecology projects in Alberta. Islamic gardens, as systems of water conveyance, are ideally suited to such production.

AGA KHAN GARDEN

↖ A modest threshold marks the entrance to the Aga Kahn Garden from the Alberta Botanic Garden. ↑ The initial encounter is the woodland *bagh* (garden), where a still pool reflects the forest and sky. The path floats over a series of wetland bowls where the forest cloisters visitors in the dense growth of the Albertan ecosystem.

↖ The design is inspired by Mughal gardens, centered on a foursquare arrangement of the *chahar bagh*. Research into horticulture, literature, and the Qur'an suggested plantings that persist in the cold climate as analogs of those found in Islamic gardens. ↑↑ The formal geometries of landscape elements invite ceremonial uses, including weddings. ↑ An amphitheater below the *talar* accommodates audiences for special events. →→ The garden's larger spatial experience of cultivation sits amid dense forests of aspen and spruce.

↖ The central *chahar bagh* is an archetypal element of Mughal gardens that emerged from agricultural landscapes. It originated in Persia. The quadrilateral layout of garden spaces is typically set within four linear watercourses. ← ↑↑ The terraced transition from the *talar* down to the *chahar bagh* is suffused with plantings specified for their fragrance, texture, and beauty. ↑ Wetland seed banks support intensive cultivation of emergent wetland species for use in restoration efforts.

↑↑ The garden's lighting at dusk illuminates voids among structural elements and reflects subtly on solid surfaces. ↑ A patterned metal screen forms an entrance threshold into the *jilau khana*, a forecourt into the garden. → A fine sheet of water spills from the plinth of the *talar* over the *chini khana*, a wall of niches that in Mughal architecture held porcelain wares, often traditionally lighted by oil lamps.

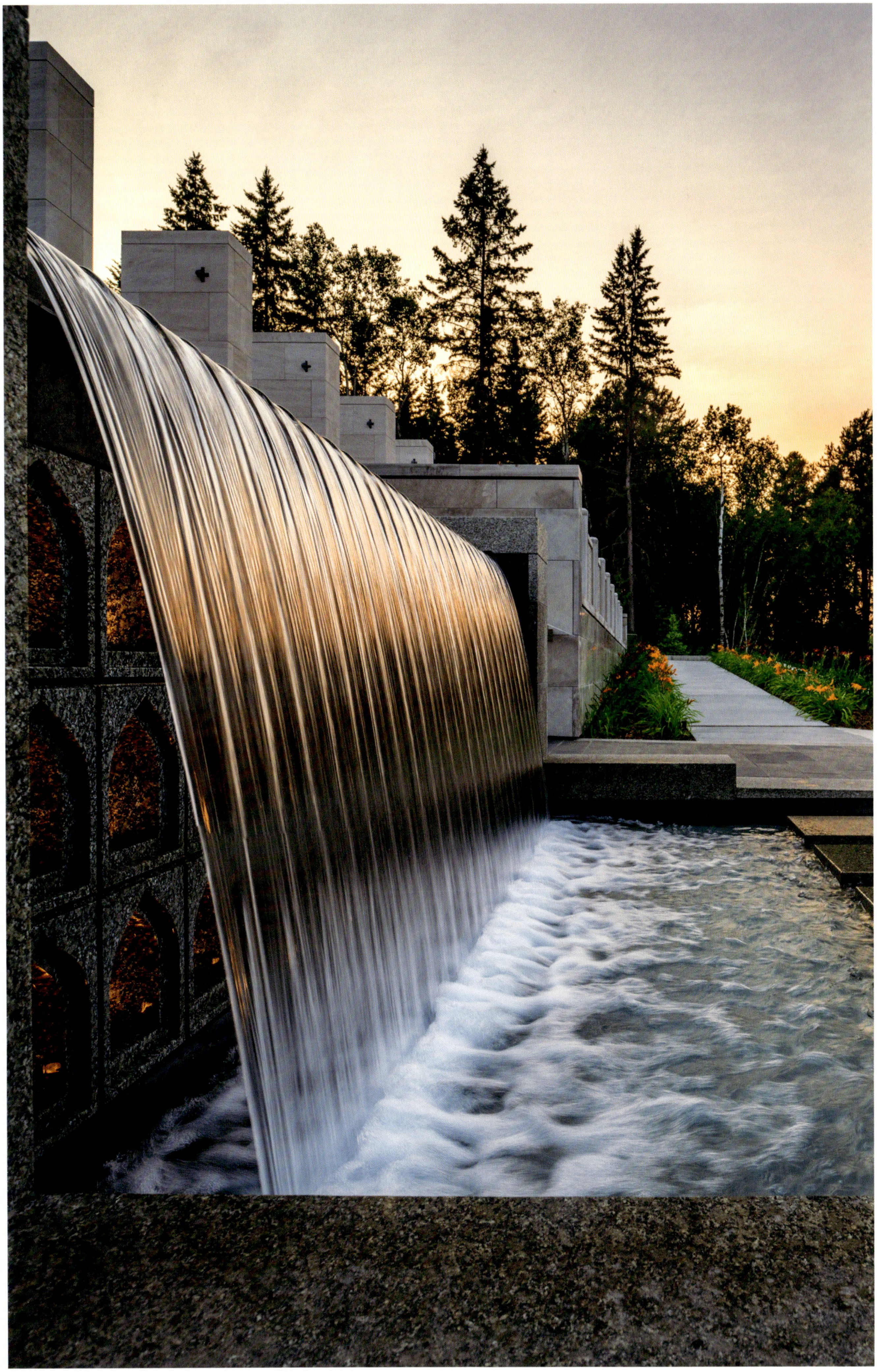

UNDIVIDED ENERGIES

PROJECT
LOCATION

MEMORIAL PARK
HOUSTON, TEXAS

The city of Houston created Memorial Park in the 1920s on land that most recently had been a military training camp. It became a forested park and eventually the largest park inside Houston's major interstate loop. By default as much as intention, the park evolved into more than a thousand acres of dense hardwoods that form a canopy as flat as the earth under Houston. Among the hardwoods, insurgent loblolly pines grow fast and punch through the canopy to spread their limbs high into the Gulf humidity. At sky level, their needled crowns float in odd equipoise amid silhouettes of high-rise buildings in every direction.

This forest, however, was in severe decline as Memorial Park neared its centenary. Many of its trees were dead or dying. Its landscapes had broken under the wear of time, neglect, and dubious planning moves. The park had limited loops of access roads, for example, but a six-lane expressway pushed through its center, cutting the grounds in half. Then in the early 2000s the forest met the force majeure of hurricanes and a devastating drought, which together destroyed half of the tree canopy. In some sections, canopy loss reached 90 percent. The forest floor held graveyards of large, shattered trunks brought to earth. Overhead, the bare limbs of the standing dead clawed at the air.

In 2012 NBW was engaged by the Memorial Park Conservancy, the city, and a group of business leaders in the nearby Uptown district to develop a long-range plan for resuscitating the park. The firm brought to the project a small college of specialists from the sciences, humanities, arts, and design to investigate Memorial Park's assets and its morbidities, to perform triage, and to reimagine its fragmented reaches as a thriving mosaic of biodiversity and recreation where people can escape the hard edges of the vast city. The plan, made public in 2015, works toward connection, reconnection, crossing synapses, making the place whole, not just on maps but among molecules and among Houstonians, who love the park as a treasure and a public trust.

The plan nests intricate, life-restoring ecological strategies among a set of larger spatial overtures. It revives biological systems that were lost or were flagging and pulls together the park's varied precincts into a more unified entity. The plan endows the park, where overlooked randomness was the rule, with clear material legibility as a place, and a place of its place.

Some of the plan's major components have now been built, among them an enormous prairie around a massive land bridge engineered to surmount the expressway that had been a mood-killing obstacle. At an opposite edge, where an orphan forest filled an eastern rump of the park, there is a new formal entrance into a lush pleasure ground called Eastern Glades set around a constructed lake. In various parts of the park, new public facilities for fitness, sports, and kids' play draw in legions of residents and visitors at all hours, and they can get around on new paths and trails that have a consistent firmness.

Major developments lie ahead: In the works is a monumental grove of memorial trees to grow in tribute to Camp Logan, the onetime training camp. The plan will also rehabilitate a large riparian zone along the southern border with Buffalo Bayou, the aorta of central Houston. In these projects like those now completed, the work of redress is drawing from deep layers of the site, from nature, history, and memory.

The plan's completed segments open immersive new experiences among spaces that had been damaged, underutilized, or just avoided. In a zone once given over to ballfields, the creation of the land bridge and the sweeping wet prairie that adjoins it transform a 140-acre area around Memorial Drive in ways that make for bracing exploration at the heart of the park. People who follow the winding paths up a gentle rise to the summit of the land bridge find the entirely novel prospect of views beyond the tree lines to expanded horizons over the city. At closer range, the mind and body register not the traffic coursing through tunnels beneath them (the traffic all but vanishes) but the potent resonance between the height of the summit and the

depth of the prairie on two sides. From the prairie's watery eye, a panorama of native meadow grasses and wildflowers sweeps and rises to cover the flanks of the land bridge. In this field of vivid colors—yellows and purples toward early summer, with brown dots of seedheads—many dozens of maples, oaks, catalpas, tupelos, sweetgums, hickories, and pecans populate the slopes, set to mature into a rolling savanna.

This landscape has a naturalistic appearance that conceals its calculation. Before the firm began programming, planning, and design of the park, it undertook a process of design research and public engagement over several months to guide appropriate changes or introductions to the park grounds. The research encompassed the area's natural history, its human habitation over thousands of years before European settlement, and its evolution into a major American metropolis with this site as a prominent public space within it.

The research phase helped the design team understand the forces through time that shaped the singular quality of the place as a microcosm of the region. The team studied the geomorphology across this area of the Western Gulf Coastal Plain; the rise of seas and formation of inland waterways such as Buffalo Bayou; the emergence of prairies and marshes that defined the regional ecology; the movements of Indigenous groups such as those of the Tonkawa, Akokisa, and Karankawa, and their relationships to the environment, including regimes of land management with fire.

Ecological research revealed vegetation types that were significant historically such as tallgrass prairie and soggy corridors of bald cypress, Southern magnolia, and post oak with minor pinelands in the mix. The firm examined the ways agriculture and extraction altered the land as a resource—loblolly pine found its home in places disrupted by timbering and plantation practices. A biological census by volunteers built an inventory of species in the park, particularly valuable in enumerating transient insect and bird populations. Reviews of soil signatures determined the types of earthen strata that defined the area, including their structure, chemistry, and drainage. The team also looked for features that merited preservation, such as important tree clusters or camp remnants. Reconnaissance of plant communities on the site, compared back to soil composition and hydrological patterns, steered the design team toward the specific vegetative communities it would aim to restore in Memorial Park along a gradient from forest to savannah to prairie to riparian corridors and edges.

The firm looked for ways the resources of the region inevitably molded human activity, back to early itinerant hunters and on to early loggers and industrialists to motorists crossing the paved matrices of modern Houston. Pre-settlement occupation trends are decoded mainly through forensic evidence on and in the land. Archaeologists for the design team canvassed for signs left behind by early European settlers, a search aided by archival sources and documents relating to the establishment of communities, industries, roads and railways, and forms of jurisdiction. The combined knowledge enabled the firm to chart a narrative timeline of human and material culture around the area that could inform and even become part of the reformation of the park.

Since the land now known as Harris County and Houston became part of Texas and the United States, the pivotal circumstance on the grounds of today's Memorial Park—the impetus later for creating the park—was the establishment in 1917 of Camp Logan, where the U.S. military trained 70,000 National Guard soldiers for combat in Europe during World War I. The camp was erected, operated, and completely taken down in less than two years, with certain traces remaining. Yet its brief presence reverberated afterward in a sense of national pride among civilians in Houston.

Some soldiers who trained at the camp and some Americans more widely would come to remember the camp as the start of the "Houston Riot," also referred to as the "Camp Logan Mutiny," and the "ugly history." In July 1917, 650 black "buffalo soldiers" of the 3d Battalion, 24th

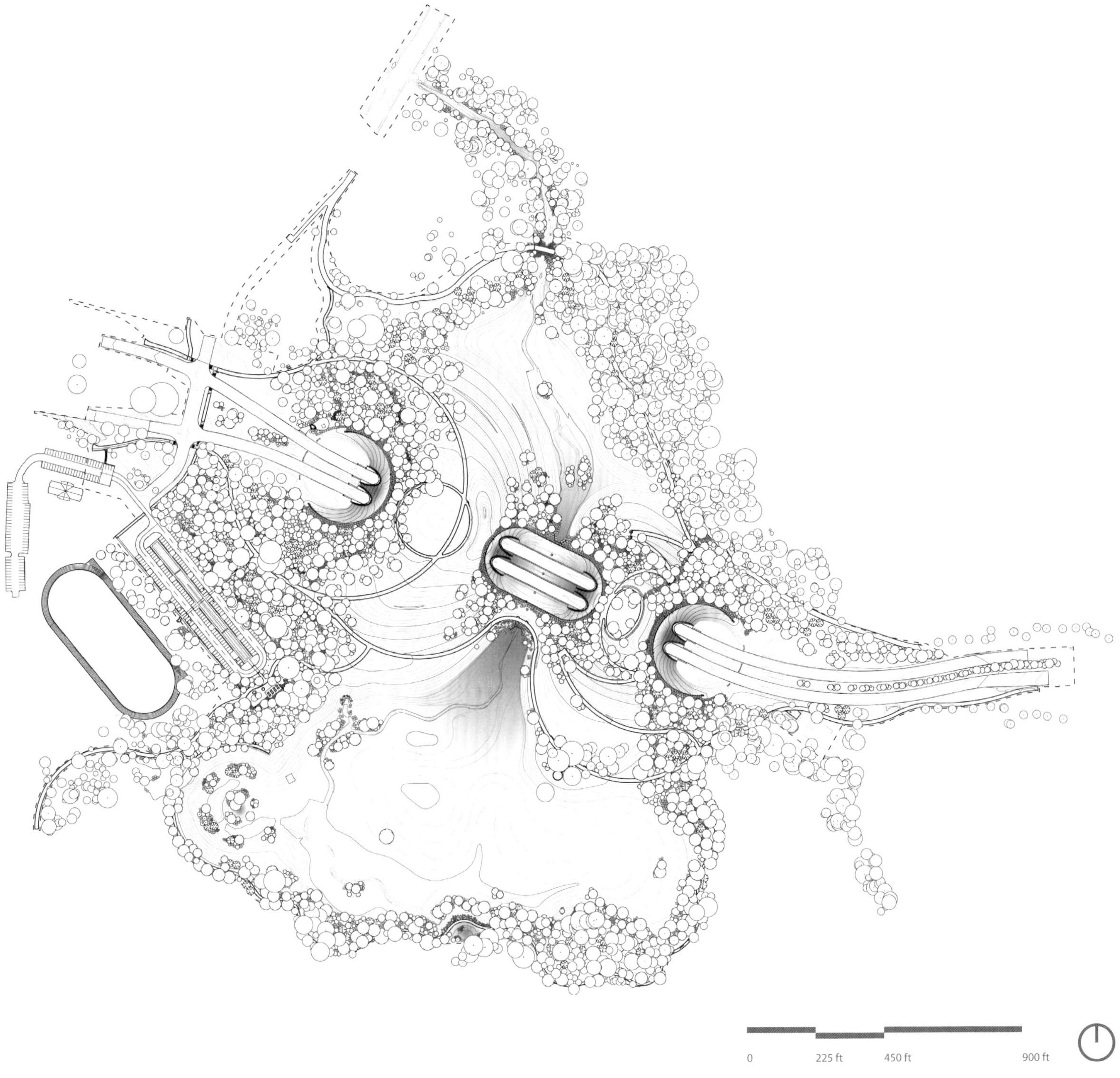

↖↑ The land bridge connects the north and south parts of the park that had long been severed by the Memorial Drive expressway. The earthen bridge also joins two sections of constructed prairie on either side of the expressway, draws a stream southward to drain into the prairie's large wet expanse, and creates a culvert for the migration of small wildlife.

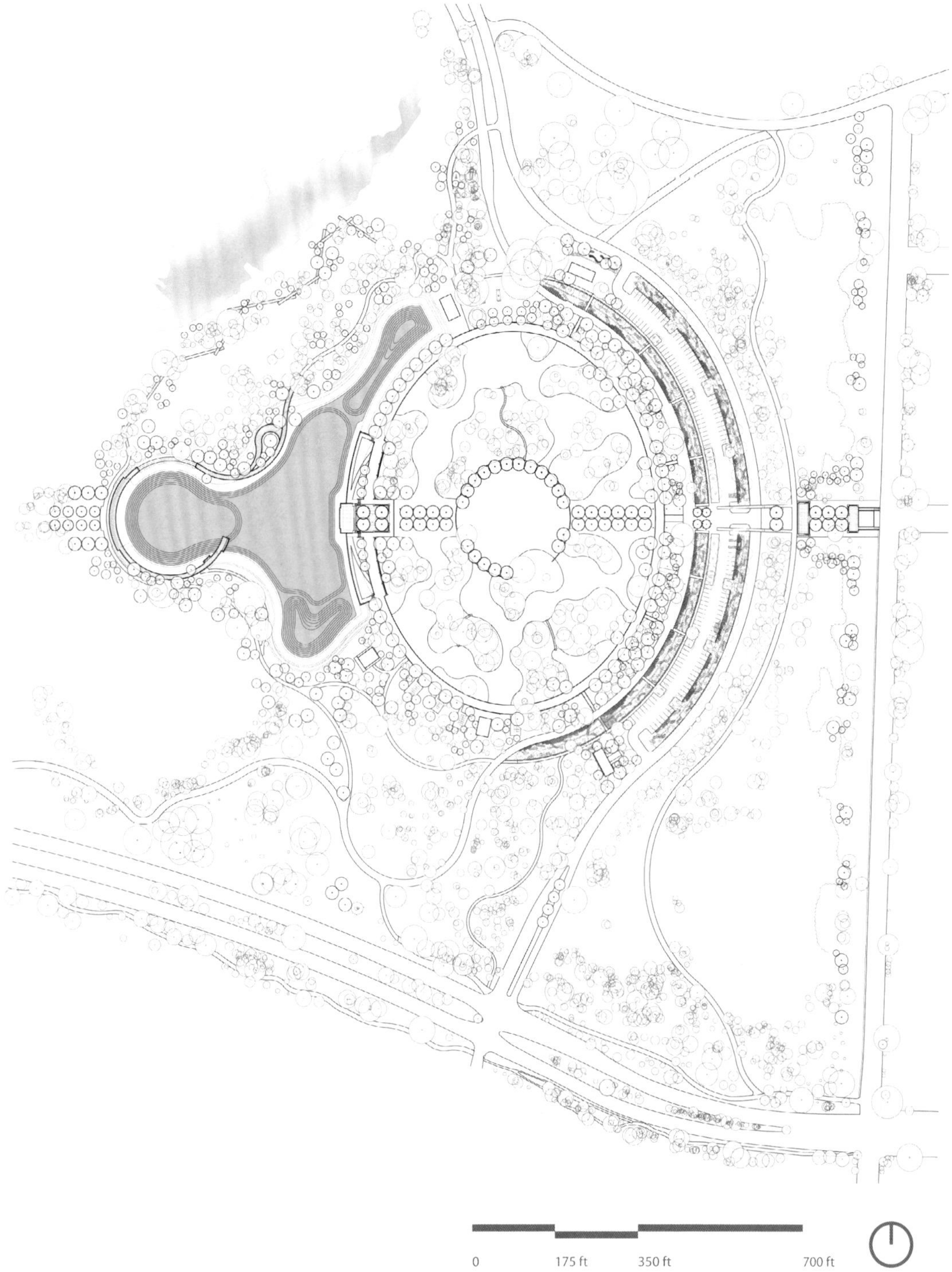

↑ Eastern Glades was carved from distressed woodland for passive recreation, picnics, and observing wildlife around the new Hines Lake, which also aids in managing stormwater. ↗ Extensive wetlands created around the five-acre lake inject thriving biodiversity to the park, which people can engage up close on sinuous boardwalks. Bald cypress are set to mature between boardwalk segments.

Infantry Regiment, were sent to guard the Camp Logan construction site under the command of seven white officers. They faced Houston's Jim Crow–era segregation and harassment by white residents and police. On August 23, mob violence and gunfire left two black soldiers and eighteen other people dead. One hundred eighteen soldiers were tried on mutiny and murder charges in courts-martial the military has since acknowledged were profoundly flawed, and 110 soldiers were convicted. Dozens received life sentences, twenty-eight received death sentences, and nineteen black defendants were executed by hanging. In 2023 the army vacated all the soldiers' convictions and posthumously granted them honorable discharges.

In the land vacuum the camp left behind, Houston's leaders wrote its afterword by creating Memorial Park. The memorial aspect about the park, like most any purposeful aspect of its grounds generally, has remained vague. In a major development phase to come, the firm's plan includes a mass planting of trees called Memorial Groves. Grids of pines will stand in regimented formation along known road alignments of the camp. They will grow tall and straight, as if at attention.

A main road into a hub of Camp Logan on the east side of the park aligned with the present-day Blossom Street, in the Crestwood neighborhood. A ghost outline of the road was found in vegetation patterns of a 100-acre tract of undeveloped land where noxious species were hastening the decline of a woodland. That tract became the Eastern Glades, the first completed major section of the firm's plan. A new park entrance, Blossom Street Plaza, opens a sequence of deep terraces grown thick with meadow plantings set within low walls—walls detailed in profiles suggestive of 1920s memorial architecture. The camp road's axis is revived in a long view to the far edge of a new five-acre lake. Boardwalks and paths encircle the lake and lead to a raised wetland pathway in the surrounding woods where bullfrogs break the still air. The heart of Eastern Glades is a nine-acre oval where the land was combed free of dead and ailing trees to make an open, round lawn at the center. Around the lawn, islets of dense growth were left to enclose smaller niches of lawn, little hideaways where people gather in small groups or alone, lolling on blankets or looking out at the lake, where black-crowned night herons slice through the air and seek prey where the land melts into water in stands of rushes, marsh grasses, irises, and water lilies.

For every apparent pleasure in these spaces—the convalescent forest; the lake a dusky orange mirror; the camplike park architecture; the deep-time narrative told along a path; lizards and wild lilies in the prairie; a new running center with a timed track—for all of those, the park holds layers of stealth physiology that support its landscapes.

The spatial arrangements of the Eastern Glades and of the land bridge and prairie are calibrated to create an aesthetically pleasing public realm. Unseen are the ways that benefit interlocks with the systemic roles these landscapes serve, not least as machines for directing heavy rainfall and filtering the pollution burdens that otherwise would have compromised larger watersheds. There is a recently constructed stream through the prairie that enables the transit of rainwater when needed from the golf course to the wet edges of Buffalo Bayou. And another facet of that solution, where the stream runs beneath Memorial Drive at the land bridge, is a sheltered culvert to let smaller land animals venture beyond the barrier of the road, broadening the reach of species' populations, which take forever to grow and no time to collapse. The park's remaking injects it with design energies meant to multiply and, given the promise of investment and close stewardship, to allow these energetic landscapes to remake themselves.

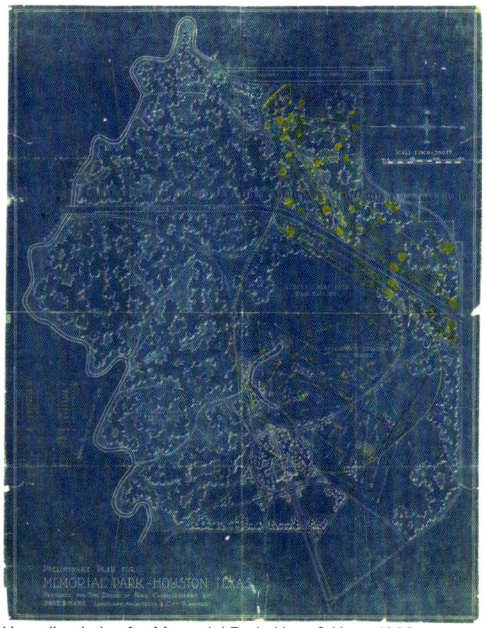

Unrealized plan for Memorial Park, Hare & Hare, 1920s.

Aerial photograph of Camp Logan, c. 1917.

Members of the 3d Battalion, 24th Infantry Regiment, detailed to guard Camp Logan construction.

The Camp Logan construction trench excavated at Memorial Park as part of a larger archaeological study.

NEWS

Drought Killing At Least Half Of Memorial Park Trees

As Houston's drought continues, one of the city's most pristine urban parks and most popular recreational areas is slowly changing. Millions of trees are dying across the city, but in Memorial Park, that problem is even more pronounced, and it could affect more than just the view.

News article describing severity of tree mortality during the 2011 drought.

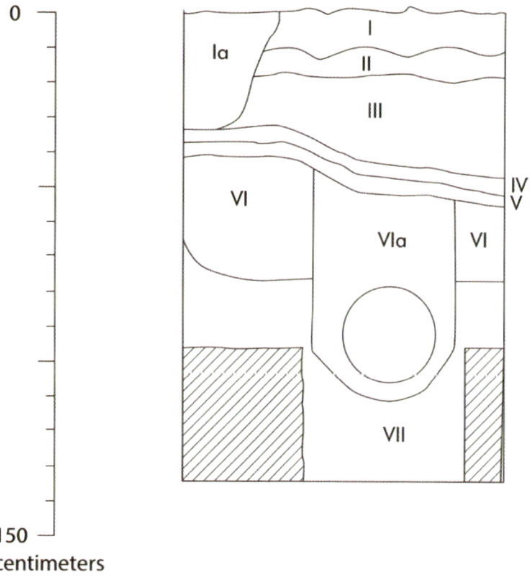

Diagram of Camp Logan construction trench excavation.

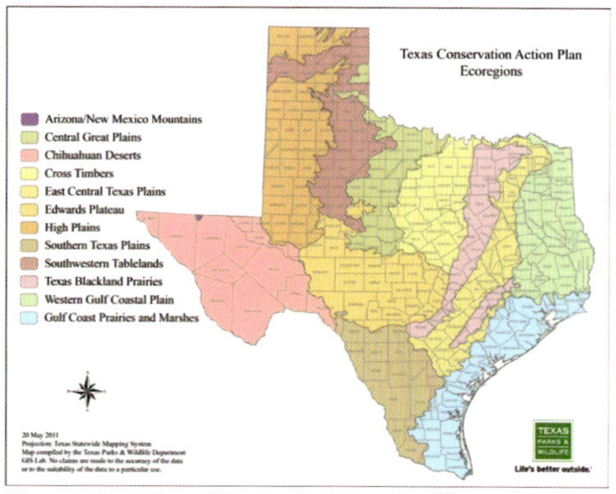

Memorial Park lies within the Gulf Coast Prairies and Marshes Ecoregion of Texas.

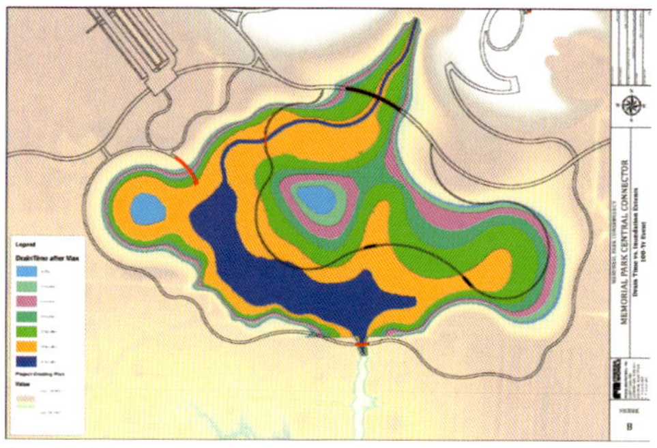
Study of the inundation capacity of the prairie during a 100-year rain event.

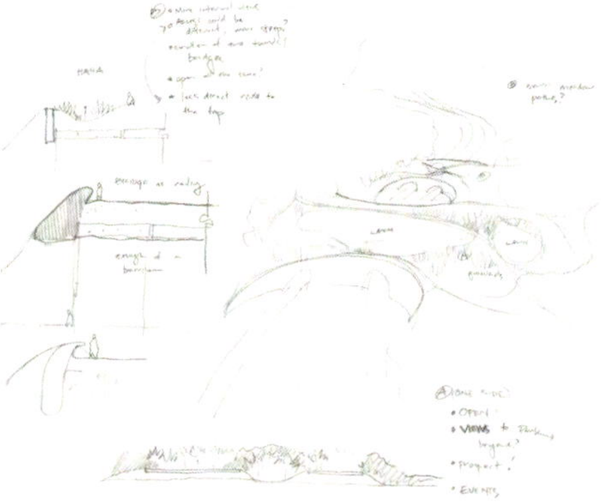
Early sketch studying land bridge function and adjacencies.

Community engagement session during early Memorial Park planning.

Soil test plots for various prairie grass mixes and exposures at Land Bridge site.

Construction of tunnels and mound for the land bridge.

↑↑ The plant species selected to populate the 45-acre prairie south of the land bridge represent many that thrived in the region before European settlement and urbanization. ↑ →→ The principal axis of Eastern Glades is a long view corridor and allée, which connects the park to the adjacent neighborhood at a new formal east portal.

↑↑ The constructed edges of Hines Lake allow for multiple ways for people to experience the water and wildlife. Threatened bird species have been returning to the area around amphitheater seating along the west edge. Lighted stone pillars conclude the principal camp-road axis of Eastern Glades that begins across the lake. ↑ A marsh of rushes and sedges fills a cove several steps from a picnic pavilion and pier.

↑↑ A pier extends into the lake from a circular path and engages the main axis of Eastern Glades. ↑ Along one edge, wide boardwalks with integral seating encourage close observation of wildlife amid emergent vegetation on all sides and lead into a trail through a forested wetland. →→ The East Terrace of Hines Lake carries the formality of the Blossom Street Plaza farther along a view axis through the Eastern Glades and brings people close to the water.

↖ A series of boardwalks leads through a forested wetland of Eastern Glades. ↑↑ The summit of the land bridge opens panoramic views of the city and sky. ↑ Precision-cut sections of the roadway removed during construction were arranged to create a challenge called the Scramble down the north slope of the land bridge. →→ The two earthworks that form the Land Bridge over Memorial Drive connect the revived prairie on the north and south and allow the stream to drain through a vale southward to Buffalo Bayou.

↑↑ The land bridge rises as an upland continuation of the prairie, from which soils were excavated to create the bridge—an intrinsic design connection. ↑ An observation pier and pavilion extends out above the prairie to allow immersion in the views and sounds of its abundant wildlife.

↑↑ The size of the prairie is calibrated to hold the enormous volume of stormwater that would accompany an event comparable to Hurricane Harvey in 2017. ↑ The prairie forms a deep, diverse, continuous carpet of native plantings up the slopes of the land bridge and between the expressway tunnel openings.

Remnant coastal prairie is scarce enough that the park's newly constructed prairie has proved immediately successful in its recruitment of a wide range of prairie wildlife, particularly bird species, both regional and migratory. The observation pier ↑↑ and two bird blinds ↑ → allow for discreet viewing of animal life at close range.

←← ↑↑ A first-rate timed track is part of a new running complex on the west edge of the park. Improved sports facilities are also found on the park's north side. ↑ New trails around the land bridge and prairie accommodate walking, running, and cycling at any speed. → The Scramble is a favorite place among children, and adults find it appealing as a strength-building exercise.

CENTENNIAL PARK
Location: Nashville, Tennessee
Date: 2013–22
Size: 132 acres

Client: Metro Nashville Parks and Recreation, Centennial Park Conservancy, You Have the Power

NBW Team: Thomas Woltz, Breck Gastinger, Alissa Diamond, Chloe Hawkins, Jennifer Trompetter, Lanie McKinnon, Andisheh Gohfranitabari, Joey Hays

Collaborators:
Phase 1: Hodgson Douglas, Sherwood Engineers, Civil Site Design Group, Princeton Hydro, BDY Environmental, Bohlin Cywinski Jackson, Centric Architecture, Silman Engineers, WiSEngineers, Hunt Design, Ashworth Environmental, SLBD Studio, Power Management Corporation, CMS Collaborative, ETM Associates, RPM Transportation Consultants, Geotek, Acoustic Dimensions, Lynch Irrigation, Thornton & Associates, Rico Associates. Connico Inc., Thomason Associates, Wilmot, Inc., Professor Scott Torreano, Natural Learning Initiatives

Phase 2: Hodgson Douglas, Civil Site Design Group, Connico Inc., Prevost Stamper Inc., Tillett Lighting Design Associates, Inc., Power Management Corporation, Rico Associates, WiSEngineers, Inc., RPM Transportation Consultants, Inc., Golf Management Group. (Children's Memory Garden) Collier Engineering, Connico Inc., Prevost Stamper Inc., Tillett Lighting Design Associates, Inc., Power Management Corporation, Rico Associates, Dowdle Construction

BOK TOWER GARDENS
Location: Lake Wales, Florida
Date: 2013–16
Size: 250 acres (total); 30 acres (NBW Design)

Client: Bok Tower Gardens

NBW Team: Thomas Woltz, Jennifer Trompetter, Nathan Foley, Sara Myhre, Sandra Nam Cioffi

Collaborators: Coyle and Caron Landscape Architects, Mary Wolf Landscape Architect, Envisors Engineers, COST of Wisconsin, The Natives, Inc., Kevin Carman

HUDSON YARDS PUBLIC SQUARE AND GARDEN
Location: New York, New York
Date: 2012–19
Size: 6.5 acres

Client: The Related Companies

NBW Team: Thomas Woltz, Serena Nelson, John Ridenour, Charlotte Barrows, Mark Strieter, Jeffrey Longhenry, Ian Brennick, Terence Fitzpatrick, Brian Schundler, Francois Poupeau, Anna Speidel, Shane Fagan, Erin Percevault, Ian Quate, Horea Popa, Maggie Hansen, Simon David, Jennifer Brooks

Design Collaborators: Kohn Pedersen Fox, Sherwood Design Engineers, Philip Habib & Associates, L'Observatoire International, Paul Cowie and Associates, Mark Smallridge and Associates, Tony Sinkosky/110CDCO Inc., Northern Designs, Pine and Swallow Environmental Services, Patrick Cullina Horticultural Design + Consulting, CMS Collaborative, Langan Engineering and Environmental Services, Vince Rico, Thornton Tomasetti

Construction Collaborators: Tishman/AECOM, General Contractor, Tutor Perini, General Contractor, Steven Dubner Landscaping, Inc., Landscape Installer, A Ottavino Corp., Stone Mason, Berardi Stone Setting, Inc., Stone Mason, Intricate Construction, Finish Concrete Installer, Situ Fabrication, West Plaza Benches and Culture Shed Bench, Urban Art Projects (UAP), Plaza Bench, Cimolai, Bollard Sheathing, Delta Fountains

BROOKLYN NAVAL CEMETERY LANDSCAPE
Location: Brooklyn, New York
Date: 2010–16
Size: 1.8 acres

Client: Brooklyn Greenway Initiative

NBW Team: Thomas Woltz, John Ridenour, Jeffrey Longhenry, Ian Quate, Margaret Hansen, Jennifer Brooks

Collaborators: Marvel Architects, Larry Weaner Landscape Associates, Grant Engineering, Jim Conti Lighting Design

DUKE UNIVERSITY RECLAMATION POND
Location: Durham, North Carolina
Date: 2010–15
Size: 12 acres

Client: Duke University

NBW Team: Warren Byrd, Thomas Woltz, Kennon Williams, Evan Grimm, Alissa Diamond, Thomas Hogge, Horea Popa, David Lepage, Kimberley Huggins, Melissa Lockwood

Collaborators: Mark Hough, Duke University Landscape Architect, Duke Facilities Management, McAdams Company

THE RAMBLE
Location: Fayetteville, Arkansas
Date: 2018–ongoing
Size: 12 acres

Client: City of Fayetteville

NBW Team: Thomas Woltz, Breck Gastinger, Jeffrey Aten, Zuzana Ponca, Nathan Foley, Joey Hays, Andisheh Ghofranitabari, Elisabeth Hoff, Christian Kochuba, Erica Mutschler-Nielsen, Kathleen Smithson, Sandra Nam Cioffi, Thomas Baker, Cong Nie, Taryn Wiens, Horea Popa, David Lepage

Collaborators: City of Fayetteville, Experience Fayetteville, Flintlock, Olsson Associates, HSA Engineering, Tillett Lighting Design Associates, Hines, Rico Associates, Archetype, Hydrodramatics, Threshold, WCRC, Nitsch Engineering, Marty Matlock, InVeritas, Nabholtz, Entegrity

ALETHIA TANNER PARK
Location: Washington, D.C.
Date: 2016–20
Size: 2.5 acres

Client: NoMA Parks Foundation

NBW Team: Thomas Woltz, Jeffrey Aten, Zuzana Ponca, Bradley Odom, Thomas Hogge, Jenn Jessup, Mandi Fung, Shane Fagan, Samantha Dabney, Chloe Hawkins, Christopher Punt, Peter Russell, Horea Popa, Heidi Petersen, David Lepage

Collaborators: Studio Twenty-Seven Architecture, Olsson & Associates, Cerami & Associates, Bowman Consulting Group, ECS Mid-Atlantic LLC, Gilmore Lighting Design, Setty & Associates International PLLC, APE Studio Inc., Robert Silman Associates, Pentagram, Irrigation Research and Design, Engenium Group, Forrester Construction

MACHICOMOCO STATE PARK
Location: Gloucester County, Virginia
Date: 2018–21
Size: 645 acres

Client: Virginia Department of Conservation and Recreation, Dominion Energy, The Conservation Fund

NBW Team: Thomas Woltz, Evan Grimm, Jen Jessup, Andisheh Ghofranitabari, Alisha Savage, Mandi Fung, Chloe Nagraj, Horea Popa, Heidi Petersen, David Lepage

Collaborators: AES Consulting Engineers, Henderson Inc., Boynton Rothschild Rowland Architects, Dunbar Structural, 2RW Consultants, Martin Gallivan, Fairfield Foundation, Gropen, Jeremy Manning, DCR, Virginia Algonquian Tribal Leaders

GEORGIA INSTITUTE OF TECHNOLOGY ECOCOMMONS
Location: Atlanta, Georgia
Date: 2018–21
Size: 8 acres

Client: Georgia Institute of Technology

NBW Team: Thomas Woltz, Lanie McKinnon, Jeffrey Longhenry, Ian Brennick

Collaborators: Georgia Institute of Technology Planning, Design, and Construction, Barge Design Solutions, Biohabitats, Long Engineering, Newcomb & Boyd, Irrigation Consultant Services, Turner Construction

FLIGHT 93 NATIONAL MEMORIAL
Location: Shanksville, Pennsylvania
Date: 2005–15
Size: 2,200 acres

Client: National Park Service

NBW Team: Thomas Woltz, Warren Byrd, Breck Gastinger, Zuzana Ponca, Paul Josey, Mary Wolf, Todd Shallenberger, Jeremy Jordan, Kennon Williams, Samantha Dabney, Fraser Stuart

Collaborators: Paul Murdoch Architects, EADS Group, RBA Partners, Sato & Bopanna, MACTEC, HF Lenz Company, Robert Silman Associates, George Sexton & Associates, Biohabitats, Louis Berger Group, US Army Corps of Engineers, Arrow Wallace Pancher, Davey Resource Group and Vanessa Hangen Brustlin

AGA KHAN GARDEN
Location: Edmonton, Alberta
Date: 2011–18
Size: 10 acres

Client: The Aga Khan Trust for Culture, University of Alberta.

NBW Team: Thomas Woltz, Breck Gastinger, Nathan Foley, Sandra Nam Cioffi, Siobhan Brooks. Jen Trompetter, Alisha Savage, Allegra Churchill, Simon David

Collaborators: Dialog Design, ISL Engineering and Land Services, CMS Collaborative Inc., Ion Irrigation Management, Inc., FTL Design Engineering Studio

MEMORIAL PARK
Location: Houston, Texas
Date: 2013–ongoing
Size: 1,465 acres

Client: Memorial Park Conservancy

NBW Team: Thomas Woltz, Jeffrey Aten, Jeremy Jordan, Bradley Odom, Katherine Tabony, Andisheh Ghofranitabari, Christian Kochuba, Thomas Baker, Taran Jensvold, Horea Popa, David Lepage, Kimberley Huggins

Collaborators:
Memorial Park Master Plan: Sherwood Design Engineers, Berg Oliver Environmental, ETM Associates, Lord Cultural Resources, DPWPR, Hunt Design, Jacob & Associates, Walter P Moore, Suzanne Turner Associates, Advanced Ecology, The Lighting Practice, Fazio Design, Lupher LLC, MIR, Hartness Vision, Camera Wings Connico IncorporatedGravity Logic, SWA, Uptown Development Authority, Houston Parks & Recreation Department, Memorial Park Conservancy

Memorial Park Eastern Glades: Uptown Development Authority, Houston Parks & Recreation Department, Walter P Moore, English + Associates Architects, Lauren Griffith Associates, Henderson Rogers Structural Engineers, G2LD Gandy^2 Lighting Design, Minor Design, Berg Oliver Associates, Advanced Ecology Limited, James Pole Irrigation Consultants, ETM Associates, Gray & Pape Heritage Management, Hunt Design, Engineers & Consultants, Inc., DPWPR, Jaime Gonzalez, Olsson Associates, James Sottilo, Professional Service Industries, Inc., Gorrondona & Associates, Richard White, Fused Industries, Landscape Art, Tellepsen Builders, Shooter and Lindsey, Pin Oak Interests, Renfrow + Co, Trees for Houston, York Bridge Solutions, Mezger, Ernst Seed, WW Bartlett , Schlitzberger Stone Designs

Land Bridge and Prairie: Ardurra, Walter P Moore, English + Associates Architects, Henderson Rogers Structural Engineers, G2LD Gandy^2 Lighting Design, Berg Oliver Associates, James Pole Irrigation Consultants, Gray & Pape Heritage Management, Hunt Engineering, Minor Design, Jaime Gonzalez, Olsson Associates, James Sottilo, Metalab Studio, Paradigm, Urban FX, Texas Project Administration Service, Uptown Development Authority, Memorial Park Conservancy, Kinder Foundation, Tellepsen, Tricon Precast, Renfrow + Co, Shooter and Lindsey, Wildlife Habitat Federation Ernst Seed, EcoServices, White Oak Studio, BCS, Texas Bomanite, Bartlett Stone Carving

Timing Track and Running Center: Gunda/Ardurra, Walter P Moore, English + Associates Architects, Henderson Rogers Structural Engineers, G2LD Gandy^2 Lighting Design, James Pole Irrigation Consultants, Gray & Pape Heritage Management, Hunt Engineering, Minor Design, Crown Castle, Pin Oak Interests, Schuam Office, PBK, A La Carte, Tellepsen, Quint Company, Texas Bomanite

NELSON BYRD WOLTZ 2024

Jeffrey Aten *
Chinar Balsaraf
Charlotte Barrows *
Paul Basque
Adeline Belsby
Ian Brennick
Thomas Cassidy
Terence Fitzpatrick
Molly Gasparre
Andisheh Ghofranitabari
Zhouqian Guo
William Hart
Joey Hays
Kimberley Huggins
Jennifer Jessup *
Kevin Kim
Michael Kittrell
Christian Kochuba
Jennifer Lauer
David Lepage *
Melissa Lockwood
Lanie McKinnon
Katherine Merhige
Hallie Morrison
Chloe Nagraj
Serena Nelson *
Bradley Odom
Alexis Perrocheau
Zuzana Ponca *
Horea Popa *
Christopher Porter
Francois Poupeau
Sandra Rice
John Ridenour *
Nathan Roth
Brian Schundler *
Kathleen Smithson
Anna Speidel
Mark Strieter *
John Tatro
Elizabeth Webster
Christopher Woods
Thomas Woltz *
Hsiao-Ying Yang

FORMER COLLEAGUES

Caitlin Ferrell
Kelly Ridenhour
Nathan Foley
Evan Grimm *
Elisabeth Hoff
June Landenburger
Christopher Punt
William Ryan
Katherine Tabony
Taryn Wiens
Robert Edwards
Ellen Garrett
Clay Gruber
Erica Mutschler-Nielsen
Cong Nie
Lesley Perez
Suzanne Hierl
Thomas Baker
Breck Gastinger *
Chloe Hawkins
Taran Jensvold
Jeffrey Longhenry *
Erin Percevault
Riley Donaldson
Jeremy Jordan *
Alykhan Neky
Ian Quate
Stephen Carlucci
Mandi Fung
Thomas Hogge
Emma Mendel
Sandra Nam Cioffi
Heidi Petersen
Alisha Savage
Samantha Dabney
Alissa Diamond
Shane Fagan
Joseph Guthrie
Kari Roynesdal
Fraser Stuart
Jennifer Trompetter
Paul Appleton
Siobhan Brooks
Matthew Seibert
Jennifer Brooks
Sara Myhre *
Margaret Hansen
Daniel Irving
Paul Josey
Kezia Ofiesh
Anne Pray
Thomas Solomon

Since January 2014: the publication date of our last monograph, *Garden, Park, Community, Farm*, and the retirement of Warren Byrd Jr.

* Acknowledging ten or more years with Nelson Byrd Woltz Landscape Architects.

THE WORK

The twelve park landscapes presented here represent a remarkable burst of creative production at NBW over the last ten years. Our community of colleagues holds the rare combination of patience and persistence required to engage in deep listening to the land and its people and the abundant design talent to envision and implement meaningful places for the public to experience. The current and former* principals of the practice merit special mention, given their long-term design leadership: Jeff Aten, Mark Strieter, John Ridenour, Jen Jessup, Lanie McKinnon, Breck Gastinger,* Evan Grimm,* Jeremy Jordan,* and Warren Byrd.*

CLIENTS AND COLLABORATORS

Publicly accessible landscapes require a particular brand of collaborative design process, and NBW thanks the staff of the institutions that steward these landscapes: Metro Nashville Parks and Recreation, Bok Tower Gardens, the Related Companies, Brooklyn Greenway Initiative, Duke University, the City of Fayetteville, NoMA Parks Foundation, Virginia Department of Conservation and Recreation, Georgia Institute of Technology, the US National Park Service, the Aga Khan Trust for Culture and the University of Alberta, Houston Parks and Recreation Department and Memorial Park Conservancy.

At NBW we know that the authenticity we seek in the public landscapes is only achieved through collaborative dialogue with communities, scientists, historians, horticulturists, engineers, archaeologists, and the creative minds that join our research-based design process. Individual contributors are listed within the preceding Project Credit pages.

THE BOOK

This book has been the focus of the NBW Communications Group over the last two years, and particular thanks goes to Horea Popa for his leadership and vision, Kimberley Huggins, production editor, David LePage, graphics editor, and Chinar Balsaraf for image research. Thanks to Elizabeth Kubany and her PR team, and special thanks to Keith Fox, Elizabeth White, and Michael Vagnetti at Phaidon/Monacelli Press for assuring a book of quality and substance, and to Damien Saatdijan, the book's designer.

THE WRITERS

A great thanks to Brad McKee for his dedication in writing the project descriptions, captions, and editing the essays. NBW is very grateful to the essayists for contributing their unique and critical perspectives of ecology, cultural history, literature, and landscape theory related to the work and design process of NBW: Nina-Marie Lister, Brent Leggs, Lawana Holland-Moore, Robert Pogue Harrison, and Andrea Wulf, the foreword author.

THE PHOTOGRAPHERS

A special word of thanks to Nick Hubbard who has worked tirelessly over the last two years to beautifully capture special moments, often on short notice, and always with a positive attitude.

Sahar Coston-Hardy, Ngoc Minh Ngo, and Max Touhey also significantly contributed to capturing the images in this book. Additional contributors include Giuliano Correia, Barrett Doherty, David LePage, David Price, and Paul Swanson.

NB

On a personal note, I want to acknowledge three individuals who continue to shape the work of NBW. Despite his retirement from practice over a decade ago, Warren Byrd and his teaching remain a major influence bridging ecology and design. Michael Bills encouraged NBW in the experimental work that led to the Conservation Agriculture Studio and twenty-five years of design dialogue and friendship. In more recent years, the presence and voice of Andrea Wulf has been a constant reminder of the magnificent interconnectedness of the planet and its people, a factor at the heart of each NBW design.

We at NBW enjoy careers of creative productivity in service to the meaningful connection between people and nature. My hope is that this book documents the talent and hearts of the individuals at NBW capturing the fruits of their personal dedication. In addition, the people who build and tend these parks are essential to sustaining the ephemeral lives of the landscapes in this book. Without the builders, users, and stewards of the future, the work will likely vanish. We express our gratitude to the next generations of stewards whom we will never know.

Thomas L. Woltz

All plans, diagrams, and project-related images are provided by Nelson Byrd Woltz Landscape Architects unless noted below. Sources are listed by row, left to right.

p. 16
Massachusetts Historical Society
Horace Cort/High Museum of Art, Atlanta
Tennessee State Library and Archives

p. 17
Tennessee State Library and Archives
University of Texas Libraries, University of Texas at Austin

p. 18
Library of Congress

p. 26
Top, left and right: Library of Congress, Geography and Map Division
Center, left: World Digital Library, Library of Congress
Center, right: Tennessee Department of Environment and Conservation.
Below, middle: National Library of Medicine
Below right: Parthenon Museum
Below, left: Chris Crookston, Wikimedia Commons
Bottom: Harry T. Burn Papers, McClung Historical Collection

p. 40
Top, left: Library of Congress, Manuscript Division, Olmsted Associates Records
Top, middle; center, left; below, middle; bottom middle: Bok Tower Gardens Photograph Collection
Top, right, center, right: John James Audubon Center at Mill Grove
Below, left: United States Department of the Interior, National Park Service, Frederick Law Olmsted National Historic Site
Center, right: Keith Novosel, Digital Content Creator, Bok Tower Gardens

p. 60
Top, left: David Rumsey Map Center, Stanford Libraries
Top, right: Markley Boyer/Mannahatta Project, Wildlife Conservation Society.
Center, left: Wikimedia Commons
Below, left: American Society of Civil Engineers
Bottom, right: The Related Companies

pp. 78, 79, 81
Nina-Marie Lister

p. 88
Top, left: Wikimedia Commons
Top, middle: Brooklyn Navy Yard Archive, Photo No. f920n1884 National Archive and Records Administration, Northeast Region-NYC, Record Group 181
Top, right: TAMS Consultants, Inc.
Center, left: Library of Congress, Geography and Map Division
Below, left: Walt Whitman, *Leaves of Grass*

p. 100
Top left: Duke University Archives photograph collection
Top, right: Mark Herboth for Affiliated Engineers, Inc.
Center, left: © 2007, The New York Times Company
Below, left: National Climatic Data Center, NOAA National Centers for Environmental Information
Bottom, right: Duke University

p. 112
Top, left: Library of Congress, Geography and Map Division
Center, left: U.S. Environmental Protection Agency (EPA)
Below, middle: Robert Ivy, *The Architecture of E. Fay Jones*

p. 130
Smithsonian National Museum of African American History and Culture

p. 131
Houston Public Library
Thomas Jefferson's Monticello.
Library of Congress, Prints & Photographs Division. LC-DIG-npcc-24262

p. 132
Nina Simone Conservancy
John Coltrane Documentary

p. 138
Top, left; center, left: Library of Congress, Geography and Map Division
Below, left: Library of Congress, Prints and Photographs Division, LC-DIG-hec-10147

p. 150
Top, left: Ashmolean Museum, Oxford, England
Top, right: Wikimedia Commons by Geographicus Rare Antique Maps
Center, left: Library of Congress Geography and Map Division
Center, middle: Henry Chandlee Forman, *Virginia Architecture in the Seventeenth Century*
Center, right: Google Earth
Below, left: Martin Gallivan.

p. 168
Top, left: U.S. Environmental Protection Agency (EPA)
Top, right: Google Earth
Center, left: Library of Congress Geography and Map Division
Center, middle: Cecil Stoughton, White House Press Office
Center, right: Special Collections and Archives, Georgia State University Library
Below, middle: Georgia Institute of Technology
Below, right: Library of Congress, Prints and Photographs Division, LC-USZ62-138800

p. 184
© blauwerke Verlag

p. 185
Epidote Press and Richard Chartier

p. 192
Top, left: Philadelphia Inquirer
Top, right: Flight 93 National Memorial/Bee and Flight 93 National Memorial/Corsa Coal
Center, bottom: Nina Bassuk
Right, bottom: Chuck Wagner

p. 206
Top, left, Center, left: © The Aga Khan Museum, AKM2, AKM3, AKM4, AKM164

p. 224
Top, left: Houston Public Library, Hare and Hare Collection
Top, middle: © Houston Public Media
Center, left: San Diego Air & Space Museum
Center, right: Uptown Houston
Center, middle: Texas Parks & Wildlife Department
Below, left: Afro American Newspapers, Gado, Getty Images.
Bottom, left, Center, middle: Gray & Pape.

THOMAS L. WOLTZ, Senior Principal of Nelson Byrd Woltz Landscape Architects (NBW), has honed a research-based design process over the last twenty-five years of practice that operates at the intersection of the culture and the ecology present in every project site. This commitment to mission and process has led the expansion of the firm to include scientists and historians as integral contributors to the design of landscapes dedicated to the sustainability of the public realm. The firm's adaptable design process is equally applicable to large urban parks, postindustrial sites, educational campuses, sites of memory, and conservation agriculture. Through this collaborative and cross-disciplinary approach, the designs reveal lost truths and erased histories in the landscape and protect them for future generations. The work of NBW now stretches across thirty states and twelve countries.

Thomas Woltz was educated at the University of Virginia and holds master's degrees in landscape architecture and architecture. He received an honorary Doctor of Science degree from the State University of New York, Environmental Science and Forestry in Syracuse. He was invested into the American Society of Landscape Architects Council of Fellows, and was named the Design Innovator of the Year by the *Wall Street Journal Magazine*. He received the Elvira Broome Doolan Medal from the Garden Clubs of America, was recognized as one of the most creative people in business by Fast Company and was granted the Land for People Award by Trust for Public Land. Woltz currently serves on the Board of Directors of the Cultural Landscape Foundation.

NINA-MARIE LISTER is Professor of Planning at Toronto Metropolitan University, where she founded and directs the Ecological Design Lab, a collaborative incubator for green infrastructure design research and practice. She was awarded the Margolese National Design for Living Prize for her work in ecological design. Lister's published research, teaching, and practice center on the relationship between landscape infrastructure and ecological systems, with applications in ecological design for climate resilience, biodiversity recovery, and well-being. Author of more than one hundred scientific articles, design essays, and book chapters, Lister is Senior Fellow of Massey College in Toronto and Visiting Professor of Landscape Architecture at Harvard University, where she directs the WildWays Project, exploring design solutions for landscape connectivity in the context of more-than-human-design.

BRENT LEGGS is the founding director of the African American Cultural Heritage Action Fund at the National Trust for Historic Preservation in Washington, D.C. Envisioned as a social movement for recognition, equity, and justice, the Action Fund empowers preservationists, entrepreneurs, artists, and civic leaders to preserve the memory and legacy of Black American identity. It is the most extensive campaign in U.S. history, with $100 million raised, 297 grantees supported, and a $14 million endowment established.

Leggs is a national leader in the U.S. preservation movement, author of *Preserving African American Places*, and recipient of the 2018 Robert G. Stanton National Preservation Award. His passion for elevating the significance of Black cultural contribution in American history is visible through his work, which elevates the remarkable stories and places that evoke centuries of Black activism, achievement, and resilience.

ROBERT POGUE HARRISON is Rosina Pierotti Professor of French and Italian Literature at Stanford University. He is the author of *Forests: The Shadow of Civilization*, a study of the role of forests in the Western imagination; *The Dominion of the Dead*, a study of burial practices and the relations that the living maintain with the dead in various secular domains; and *Gardens: An Essay on the Human Condition*, among others. In 2014 he was knighted Chevalier by the French Republic, and he is a member of the American Academy of Arts and Sciences and the host of the podcast *Entitled Opinions*.

ANDREA WULF is an award-winning author of seven acclaimed books, including *Founding Gardeners* and *The Invention of Nature*, and most recently, *Magnificent Rebels: The First Romantics and the Invention of the Self*. She has written for the *New York Times*, the *Atlantic*, the *LA Times*, the *Wall Street Journal*, the *Guardian*, and many other national and international publications. She has lectured worldwide—from the Royal Geographical Society and Royal Society in London to Monticello and the New York Public Library.

BRADFORD MCKEE writes about environments, design, and culture and was editor in chief of *Landscape Architecture Magazine* from 2010 to 2020. Previously, he was a senior editor of *Architecture* magazine and the arts editor of the weekly *Washington City Paper*. He has also worked as a contract reporter for the *New York Times* and was a regular contributor to the magazines *I.D.* and *Architect*.

PHOTOGRAPHY CREDITS

Collina Strada 95

© Sahar Coston-Hardy/Esto 5 (top left), 135, 139, 140-141, 142, 143, 144, 145 (bottom)

© Barrett Doherty 4 (middle left), 37, 58, 61 (bottom), 63, 64–65

Cameron Franchuk, six mile road photography 216 (bottom)

The Greathouse Company, LLC 27 (top)

Nick Hubbard 4 (top left, middle right, bottom left, bottom right), 5 (middle left, bottom left, bottom right), 23, 24, 30–31, 32 (bottom), 33, 34 (top), 38, 41, 42–43, 44, 45, 46–47, 48, 49 (bottom), 50, 51, 52–53, 54, 57, 61 (top), 62, 66, 67, 68-69, 70, 71, 97, 98, 101, 102-103, 104, 105, 106, 107, 109, 110, 113, 114–115, 116, 117, 118–119, 120, 121, 122–123, 147, 151, 152–153, 154, 155 (top), 156, 157, 158,159, 160–161, 162, 163, 165, 166, 169, 170–171, 172, 173, 174–175, 176, 177, 220, 223, 225, 226–227, 228, 229, 230–231, 232, 233, 234–235, 236, 237, 238, 239, 240–241, 242, 243, 244–245

Jacq Justice 28 (bottom)

Sam Kittner 136

David Lepage 155 (bottom)

© Ngoc Minh Ngo 27 (bottom), 28 (top), 32 (top), 86, 90, 91, 94

NoMa BID 145 (top)

Keith Novosel 55

David Price 49 (top)

Eric Staudenmaier 190

Trey Stephens 219

Sterling E. Stevens 34 (bottom), 35

Paul Swanson 204, 208, 214 (bottom)

Max Touhey 4 (top right), 5 (top right, middle right), 85, 89, 92-93, 189, 193, 194–195, 196, 197, 198–199, 200, 201, 203, 207, 209, 210, 211 (top), 212–213, 214 (top), 215, 216 (top)

Jeff Wallace 211 (bottom), 217

Virginia Department of Conservation and Recreation 148

Nathan Zucker, courtesy of Centennial Park Conservancy 29

COLOPHON

First published in the United States by The Monacelli Press.

Library of Congress Control Number: 2024940642
ISBN: 978-158093-660-6

Design: Damien Saatdjian

Printed in China

Monacelli
A Phaidon Company
111 Broadway
New York, New York 10006